Active Science Three

Albert James B.Sc.

Illustrated by Edward Taylor

SCHOFIELD & SIMS LTD HUDDERSFIELD

© 1978 Schofield & Sims Ltd

All rights reserved.
No part of this publication may be reproduced,
stored in a retrieval system, or transmitted,
in any form or by any means, electronic,
mechanical, photocopying, recording or otherwise,
without the prior permission of Schofield & Sims Ltd.

0 7217 3545 2

First printed 1978
Second impression 1978
Third impression 1979
Fourth impression 1980
Fifth impression 1981

Active Science is a series of four books:

Book 1	0 7217 3543 6
Book 2	0 7217 3544 4
Book 3	0 7217 3545 2
Book 4	0 7217 3546 0

Acknowledgements

The author and publishers wish to thank the following
for permission to use copyright photographs:

The Trustees of the British Museum: pp. 5 (upper), 6 (upper)
The National Museum of Antiquities of Scotland: p. 5 (lower)
Reed Engineering and Development Services Ltd.: p. 10 (2)
International Wool Secretariat: pp. 11 (2), 12 (2), 13 (centre)
Shirley Institute, Manchester: p. 13 (left and right)
Cement and Concrete Association: p. 15 (upper)
Daily Telegraph Colour Library: pp. 15 (lower), 37 (lower)
Keep Britain Tidy Group: p. 18 (upper and lower)
Abbey Car Breakers: p. 18 (centre)
ICI Hyde Group, Paints Division: p. 20
Doug Kincaid, Esq.: p. 22
British Steel Corporation: pp. 26 (upper and lower), 44
Bertie Daish, Esq.: p. 32
Wates Ltd.: p. 37 (upper)
Grattan Warehouses Ltd.: p. 41
Bruce Coleman Ltd.: pp. 48, 74

Printed in England by Chorley & Pickersgill Ltd Leeds

Contents

1 Materials we use

Metals

How many objects made of metal can you see in the picture?

Make a list of the names of all the different metals you know.
What metals are used for making kitchen pans?
What metals are used for wires to carry electricity?
What metals are used for making jewellery?
What is the metal used for car bodies?
What is the metal used for the shiny plating on car bumpers?
What metal is used in car batteries?
What two metals can you find in a torch battery?

Alloys

Many of the metal objects you will see are made from an *alloy*. To make an alloy, two or more metals are melted together in the right proportions, mixed well and then allowed to cool and become solid again.

Sometimes a non-metal like carbon is included.

Bronze

The first alloy to be made was *bronze*. Men had learned how to get copper by heating certain rocks. Copper was found to be rather a soft metal, and it could not very well be made into

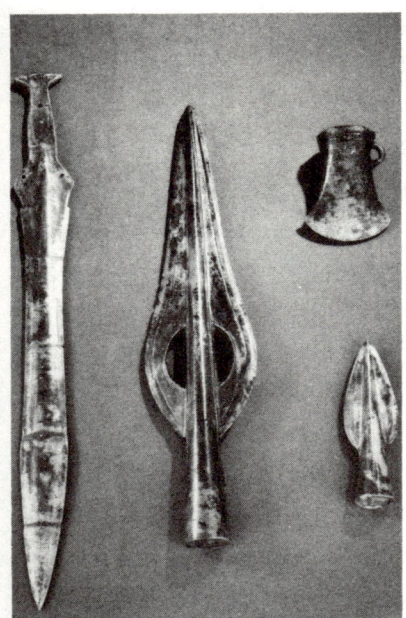
Bronze Age weapons and tools

tools, or given a sharp edge. The metal tin was also known, but it was just as soft as copper. Then it was discovered that if copper was melted with a small amount of tin, the alloy cooled to a *hard* metal quite different from either of the original metals. The alloy was called bronze. It was harder, stronger and tougher than either copper or tin and would sharpen to make axes, spears and knives. It made better tools than any material known before. The discovery of bronze began a whole new age of progress and development for human beings, called 'The Bronze Age'.

Brass and steel
Brass is an alloy of copper and zinc.
Steel is the alloy that we use most of all. It is an alloy of the metal iron with the non-metal carbon. By varying the amount of carbon used, and also by adding small amounts of other metals such as manganese, chromium, cobalt, nickel or tungsten, many different kinds of steels are produced. Some steels are stainless, some are specially hard for making drills, some are specially strong for the cables of suspension bridges, some make good magnets and so on.

Coins
Coins used to be made of pure gold, silver or copper. The value of the coin used to be the actual value of the metal it was made from. Now, coins are just tokens and the metal in them has very little value. They are usually made from alloys of copper and nickel.

A James III silver coin, called a groat (Scotland 1485)

Object	Metal or Alloy it is made from
saucepan	aluminium or iron
nail	iron
water-pipe	copper
spring	steel
cooking-foil	aluminium
tin can	iron covered with a thin layer of tin
soft drink can	aluminium alloy
car body	steel
torch battery case	zinc
ornaments	brass or copper
aeroplane fuselage	aluminium alloy

Metals and alloys we use

The table shows a list of some metal objects with the name of the metal or alloy each is made from.
Make a copy of this list. Then see if you can add to the list.

The properties of metals (what metals are like)

See how many different metals and alloys can be collected by your class.

1 Metals are shiny
With emery-paper, clean some of the samples you have collected and polish them with metal polish so that they shine and the true colour of the metal can be seen.

A bronze mirror made in Britain in the first century A.D.

Bright chromium-plated reflector in car fog-lamp

Metals go dull when left in the air, especially if it is damp. We say that they *tarnish*. Some metals tarnish quickly, some slowly. Put polished pieces of different metals outside. Inspect them every day for a few weeks. Write down what happens. Roman ladies had mirrors made from flat pieces of silver or bronze. They had to keep them well polished. Now we have metals such as chromium and nickel which tarnish very slowly. As these metals keep bright for a long time without polishing, they are useful for such things as car headlight reflectors.

2 Weight

Do all metals have the same weight?

Lift up some pieces of different metals of about the same size. What do you notice about lead and aluminium, for example? Perhaps your teacher can help you to get pieces of different metals exactly the same size. Weigh each piece and arrange them in order, from the heaviest to the lightest.

3 Bending

Ask your teacher to help you to obtain a number of different metals (e.g. iron, copper, brass, aluminium, zinc, lead) in the form of strips. Each strip must have the same length, width and thickness. Test how well they bend, as shown in the picture.

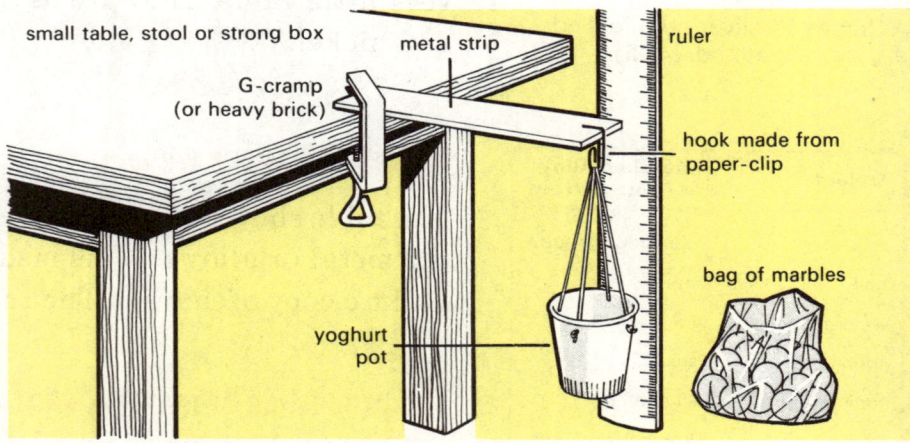

small table, stool or strong box   metal strip   ruler

G-cramp (or heavy brick)

hook made from paper-clip

bag of marbles

yoghurt pot

Always make sure that you have the same length of strip for each test, measuring from the table edge to the end. Always start with the strip quite level and then measure how much each strip bends for 1 marble-weight, 2 marbles-weight, and so on.

Make a chart of the results like this:

Metal	How far the strip bends with the weight of:					
	1 marble	2 marbles	3 marbles	4 marbles	5 marbles	etc.
zinc						
iron						
brass						
copper						
lead						
aluminium						
etc.						

4 Metals and magnets

Some people say 'magnets attract metals'. Find out if this is true by testing some metals. Test iron, copper, steel, aluminium and any other metals you have. Do you think magnets attract all metals?

5 Hardness

Can you find out which of the metals you have is the hardest? Try to make a notch in each metal with the edge of a file. Judge how easy it is to do this to each metal. Compare your findings with those of other people in your class to see if they think the same as you do.

A better way to compare the hardness of different metals is to try scratching each kind of metal with the sharp corner of another kind of metal. If one metal will scratch another, it must be the harder of the two. Use a magnifying glass to be sure you *are* making a scratch and not just marking the surface. The simplest way of showing your results is to arrange a row of metals and alloys in order of hardness.

Wood

Different kinds of wood

Different types of trees give different kinds of wood. See how many kinds of wood you can collect. Look at the grain and the colour of each type of wood. Note how the different woods smell when they are freshly cut. Sandpaper and polish your different pieces of wood and see how they look then.

Ask your teacher to help you to get some blocks of different woods all cut to the same size (a good size would be 20 cm × 10 cm × 2 cm).

Do all the equal-sized blocks weigh the same? Arrange them in order of weight.

The grain pattern on a piece of wood

Uses of wood

Wood is one of the most useful materials we have. How many ways of using wood can you see in the picture?

What other uses for wood do you know?

Collecting wooden objects

1 You can collect beautiful wooden things. For example, a smooth wooden bowl, a carving and a piece of driftwood from the sea-shore.
2 You can make wood carvings yourself. It is possible to carve quite well with a sharp penknife. Try to find which kind of wood you like best for carving.
3 You could make a collection to show the different kinds of wood used in sports and games.
   What kinds of wood are used for making furniture or for making musical instruments? Make a collection of photographs of wooden furniture and musical instruments.

4 Make a collection of 'man-made' woods. Plywood, hardboard, insulating board, chipboard and blockboard are examples of man-made woods. How many more can you find?

Floating wood

How well do different kinds of wood float? You can tell by putting pieces in water and looking how much of each one is under water and how much is out of water. This is not easy to estimate with some pieces because they float in odd positions. If you can cut pieces shaped like the one shown in the drawing, you should find it easier. Blocks of this shape always float in the position shown and if you put them in water together, you can judge the differences between them quite well.

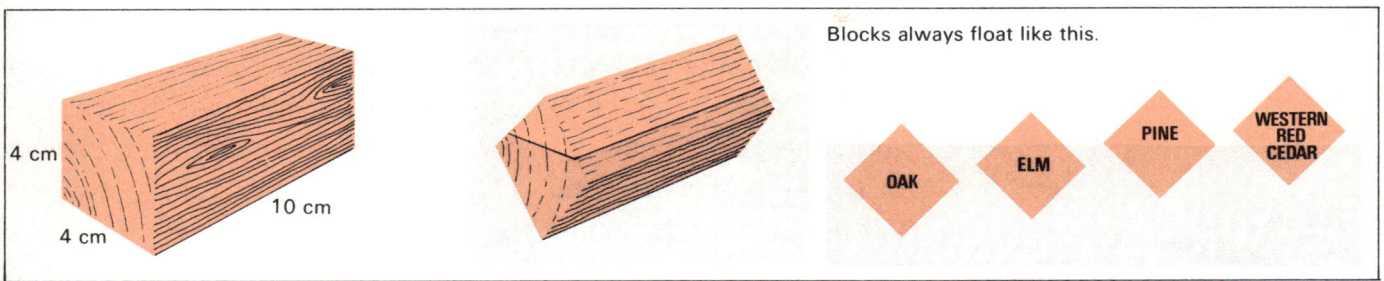

Blocks always float like this.

4 cm
4 cm
10 cm

OAK ELM PINE WESTERN RED CEDAR

Leave the blocks in the water for a day or two and then see how they float. Do they float differently from before? Which of all the woods you have would make the best raft?

If you can get a piece of ebony or a piece of the wood called lignum vitae you will find that these woods sink in water.

Hardness

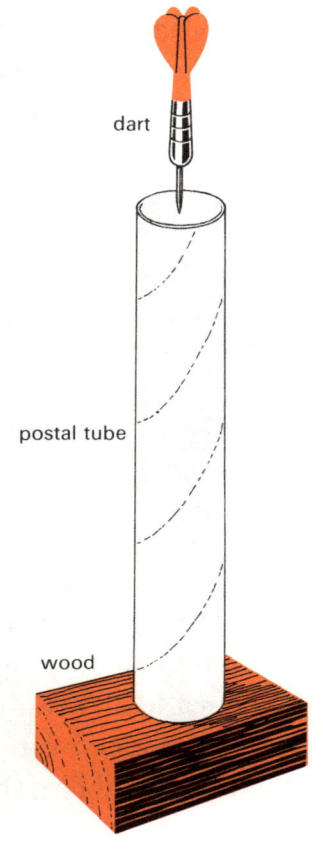

dart

postal tube

wood

If you try to do some carving, you will find that some woods are a good deal harder than others. There are many other ways of testing the hardness of woods. The drawing shows one way.

The cardboard tube makes the experiment safer and also makes sure that the dart drops from the same height each time. You have to measure how deeply the dart goes into each different kind of wood. What do you think is the best way to do this?

Do not be satisfied with just one drop of the dart for each kind of wood. Wood is not of the same hardness all over, so you might hit a hard spot or a soft spot. How many goes do you think would be fair for each piece of wood? Can you find an average of your results? Use this method to arrange a set of different kinds of wood in order of hardness.

Think of another method for testing the hardness of woods — one of your own ideas. See if you get the same results with your method as you did when using the dart.

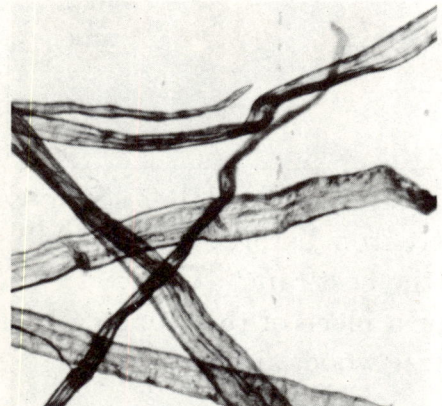

Newsprint (× 50)

Wood-pulp fibre (× 120)

Paper

See how many different sorts of paper and card you can collect. Make an exhibition of them.

Look at pieces of paper with a magnifying glass or a microscope. Look for strands and fibres. It is often easier to see these at a torn edge.

Paper is made up of fibres.

These fibres usually come from wood. To make paper, pieces of wood are shredded and smashed by forcing them against a 'grindstone' in a large amount of water. This makes a pulp of wood fibres which is forced through bars to cut the fibres to a smaller size. After being treated with chemicals in a large tank, the pulp is partly dried on a fine sieve and then passed on to the paper-machine. This machine has a long conveyor belt made of finely woven wire-cloth which rests on rollers. As the rollers turn they move the belt along and as it moves the pulp is poured over it. A great deal of the water drains away through the wire-cloth. Finally, the mass of fibres left on the belt is pressed between large rollers which squeeze it out into a thin sheet and then this goes over steam-heated rollers to dry it.

Instead of using the rollers, some paper is hand-made by pouring out pulp in a thin, even layer on a flat tray with many fine holes in the bottom. There it is simply left to drain and dry to make a sheet of paper. You might like to try to make your own paper but it is not easy. You may not be able to pulp wood, so try other things. Some children have done quite well with cow-parsley and others have made paper from plantain leaves. The very first paper was probably made in this way from riverside reeds.

Papier mâché

You will enjoy making papier mâché. Tear up old newspaper or tissues into small pieces. Soak them in water with a little wallpaper paste added and stir. When you get a good pulpy mass, squeeze and mould it to the shape you want. It will dry hard like wood after several days, and then you can paint it. Why not make some heads for puppets like those shown in the picture?

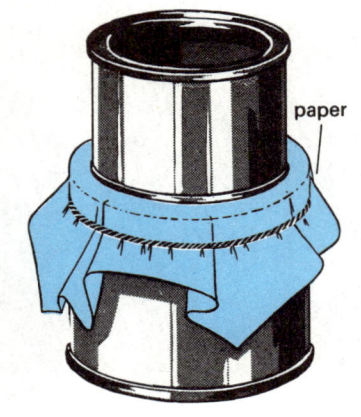

paper

Strength of paper

Some paper tissues are supposed to have 'wet strength'. Here is one way to test the strength of paper tissues, both when they are dry and when they are wet.

You will need two small cans, or other containers, of the same shape. One should be smaller to allow it to slide easily into the other. The piece of paper to be tested is tied firmly over the top of the bigger can. The smaller can is placed on top of the paper and loaded with marbles, washers or other small weights, one at a time, until the paper tears. Do the experiment with dry tissues and then with wet ones.

Compare the number of weights needed to tear dry and wet paper tissues.

Try several makes of paper tissues to see if one make is better than the others.

Compare the strengths of many other kinds of papers besides tissues. Make a chart to show the results.

There are other ways to test the strength of paper. Invent one for yourself. Does it give the same results as the one described here?

Cloth

Fibres used to make cloth

Find pieces of cloth made from wool, cotton, silk, flax and man-made fibres. (The cloth made from flax is called linen.) Which of these materials came from plants? Which came from animals? What is the starting material for making man-made fibres?

Weaves and patterns

Knitted cloth

Woven cloth
How the threads are woven together

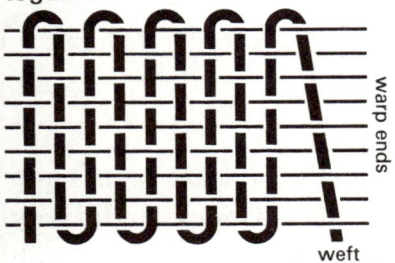

warp ends

weft

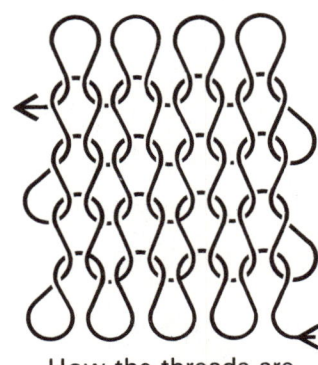

How the threads are knitted together

A very simple woven pattern
two under – two over makes twill ⟶

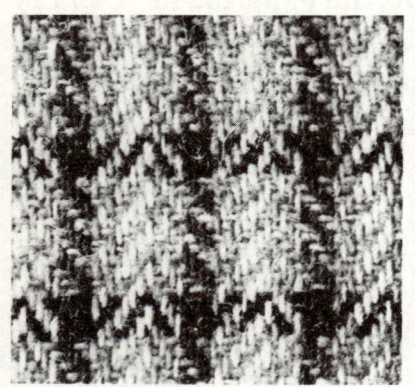

A more complicated woven pattern

Make a collection of as many different fabrics as you can. Look at your different fabrics with a magnifying glass. What do you notice? Can you tell how the fabrics were made? Some fabrics are woven and some are knitted.

The pictures on page 11 show you the difference between woven and knitted fabrics.

Look at your different fabrics again.

Can you see how the patterns have been woven into the cloths?

How to recognise different fibres

If you do not know what a piece of cloth is made from, there is a good way of finding out. To find out how the test works, first get some threads which you *know* to be cotton, wool, silk, linen and one of the man-made fibres, say nylon.

Hold one end of each thread in turn in a candle flame.

For each thread, write down

1 how it burns;

2 what the burning thread smells like;

3 what is left after burning.

Make a chart like this.

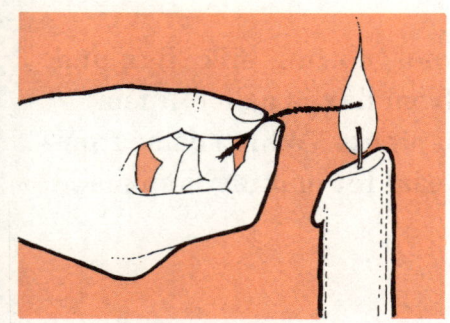

Thread	How it burns	Smell	Remains
Cotton			
Wool			
Silk			
Nylon			
Linen			

Now get some different pieces of cloth. Pull a fibre from each piece of cloth and test it to see how it burns. Compare the way it burns with the table you have made. Can you tell which fabric each of your pieces of cloth is made from? *Remember, for safety you must burn only ONE thread at a time and always take great care with the flame.* Be sure to put it out when you have finished. The next pictures show what different fibres look like under a microscope.

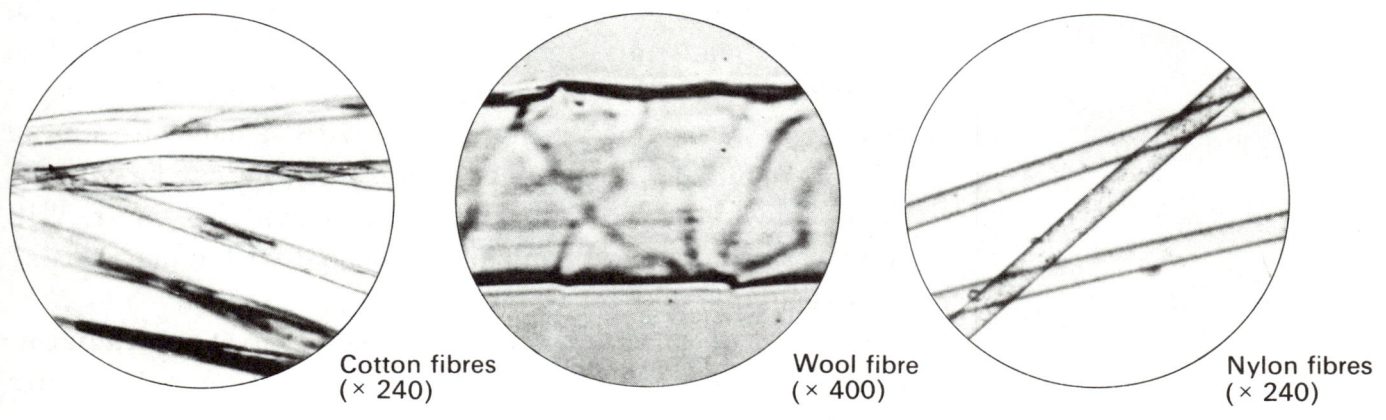

Cotton fibres
(× 240)

Wool fibre
(× 400)

Nylon fibres
(× 240)

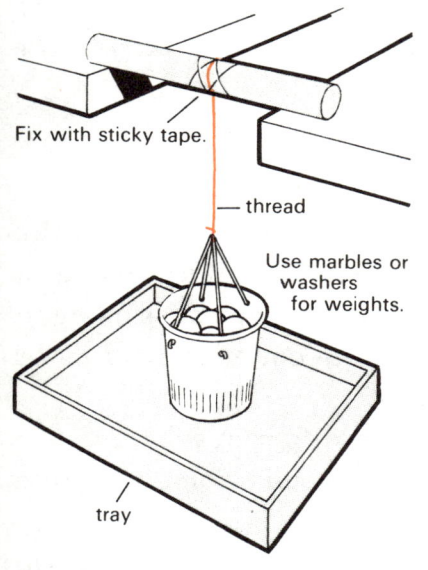

Fix with sticky tape.

— thread

Use marbles or washers for weights.

tray

Strength of threads

The drawing shows you how to test the strength of a thread. How strong are different types of threads? Test sewing threads and knitting wool. Try to be fair and use a single thread. You may have to do some untwisting to get a single thread. Try also to get threads which are as nearly as possible all of the same thickness. Pull threads from different kinds of cloth and test them. Do not forget to do each test several times with each kind of thread to see if you get the same result every time. What will you have to do if you get different results with the same kind of thread?

Find the strongest thread you have, and the weakest.

Plastics

1 How many things made from plastics are shown in the next picture?

2 Make a list of all the things made from plastics which you use and notice in a day. You will probably be surprised how long it is.

3 Make a collection of plastic things for the class-room. Arrange it to show the different kinds of plastic materials there are. Some are hard, some are soft, some bend easily, some are rigid. There are many different colours and textures. Some feel waxy, some smooth, some rough.

4 Test some scrap pieces of plastics with a penknife or old scissors to find out how well the different kinds can be cut.

5 Find out whether all plastic materials float on water. Will your test be fair if you compare a toy boat with a plastic tile? How will you make it fair?

6 Put pieces of different kinds of plastic on a hot radiator. What happens? Cool the same pieces in a freezer. What are they like now?

Keep plastics away from flame; some are flammable. Even those which simply melt in a flame can give you a nasty burn if the molten material falls on your skin.

Concrete

Concrete is made by mixing gravel, sand and cement in the right proportions with water. Watch a builder using a concrete-mixer. Try to find out what proportions he is using in the mix. How is he measuring? Does he count shovelfuls or does he measure in some kind of box?

Here are some of the mixtures used for different jobs.

JOB	PARTS		
	cement	sand	gravel
Foundations	1	3	5
Floors and beams	1	2	4
Roofs and tunnels	1	2	3

Get some cement, sand and gravel and mix some concrete yourself.

Be sure to use only enough water to make a stiff mixture. Do not make the mixture too wet.

Make some moulds like those in the picture and fill them with the concrete mixture.

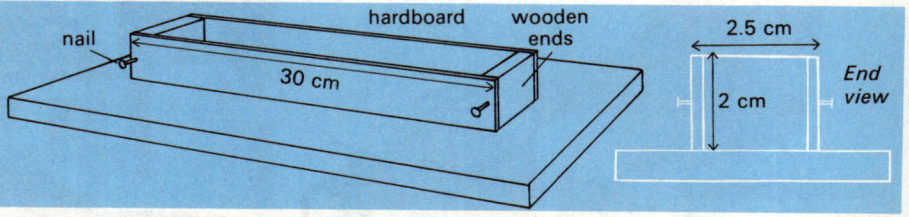

Leave the concrete for several days to set hard. Then take away the mould. The next picture on page 15 shows you how to test the strength of your concrete bar.

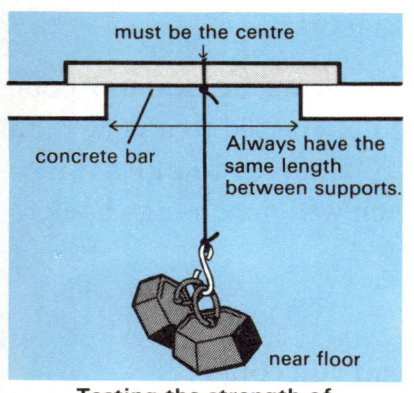

must be the centre

concrete bar

Always have the same length between supports.

near floor

Testing the strength of a concrete bar

Make some more bars exactly the same size but with different proportions of cement, sand and gravel in the concrete mixture. Be sure to label them carefully because they will all *look* very much the same. Test the strength of each bar.

Which is the strongest bar?

Have you seen builders using *reinforced* concrete? This is concrete with a network of steel rods or wires set inside it.

Building a bridge with reinforced concrete

Make a concrete strip in a mould exactly as you did before, but this time put one or two iron or steel rods, wires or strips inside the wet concrete mixture.

When the concrete has set, test the strength of the strip as you did the others. What do you notice?

Clay

Wet and dry clay

1 Mould some shapes out of clay and dry them in the sun, over a radiator, or in an oven. The clay will go quite hard, but if you wet it with water it will slowly return to soft clay again.

2 Clay shrinks as it dries. You can see how much by doing the experiment shown in the picture.

Dry clay shrinks and cracks

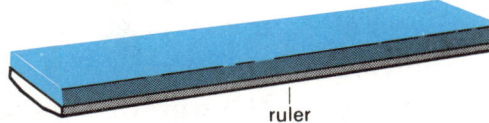

Cut block of wet clay about 10 mm thick so that it fits ruler exactly.

ruler

What happens when it dries?

Have you noticed large cracks in lawns on a clay soil in summertime? What do you think has caused these?

Baked clay

Clay is a most useful material for making bricks and pots. To make bricks and pots, clay has to be heated to a very high temperature in a *kiln*. Then the clay changes to quite a new material which is very hard and which will *not* change back to clay again if you put water on it.

Look at bricks

How many different kinds of brick can you collect?
Are they all the same weight?
Are they all the same size?
What different colours are there?
What different surface textures can you find? Make some texture-prints from the most interesting ones. To do this, hold a piece of paper quite firmly on the brick and then rub over the top of it with a wax crayon.
What words do you find moulded into the bricks? Can you tell where the bricks were made? Find the places on a map.

Bricks and water

1 How much water will a brick soak up? Weigh the brick dry, then soak it for half an hour in a bucket of water and weigh it again.

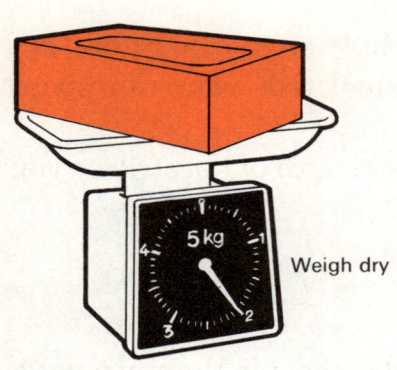

Weigh dry

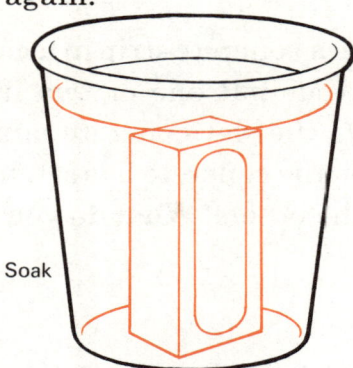

Soak

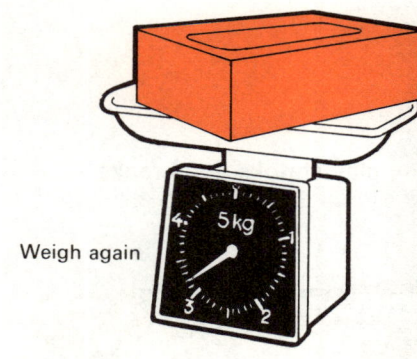

Weigh again

What weight of water has it absorbed? Try different kinds of bricks. Does each soak up the same amount of water? Try a blue (engineering) brick. What do you find out about this one?

2 Stand a clean, dry brick on two pencils in a dish of water as shown in the picture.
Can you see the water rising up the brick?
How long does it take to rise 2 cm?
How far has it risen in a week?

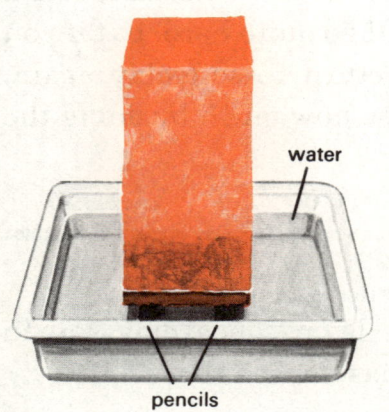

water

pencils

House walls being built

Try several different kinds of bricks. Does the water rise at the same rate in all of them?

Look at a house and find where the *damp-proof course* is. What is it made of? What would happen to the walls of the house if it was not there?

Building walls

The next picture shows house walls being built. You can see two layers of bricks with a cavity between them. Find out how wide the cavity is usually made. What do you think it is for? Why does the builder use a different kind of brick for the inside walls?

There are several ways of fitting bricks together to build a wall. Can you find walls built like those shown in the pictures? Can you find any others?

Flemish bond

English bond

Glass

Glass is made by melting sand. It needs a very hot furnace to do this. To make the glass clear and to produce different kinds of glass, small amounts of different chemicals are melted with the sand. For ordinary glass, soda and chalk are often used. How many different uses for glass do you know?

A lot of rubbish is dumped on local tips

A car crushing machine

Litter left by roadside

Throwing materials away

We throw away so much waste material that it is difficult to know where to put it all.

Find out how many different kinds of material are thrown into your dustbin in a week. When your dustbin is emptied, where is the rubbish taken to? Is it put on a dump or is it burned?

It would be interesting to measure how much waste comes from your school in one day. It should be fairly easy to have it all put into large plastic bags. Then you could weigh it. Work out approximately how much waste comes from your school in a week.

Does your local council salvage anything from the rubbish they collect? What are they going to do when the rubbish dumps they use now are full up?

The disposal of old cars is a problem. The picture shows a machine which will squash an old car to a fraction of its former size. Then the steel can be melted out to be used again.

Litter

Some people are careless and leave a great deal of untidy litter around.

Suppose you drop some toffee papers on the ground. How long do you think it takes for them to rot away?

Try an experiment. Fix some toffee papers to the ground with stones or wooden pegs so that they do not blow away. Look at them from day to day and record what happens to them.

Try the same experiment with newspaper, writing-paper, paper tissues, cardboard and a plastic bag.

Leave outside a tin can and a plastic bottle. How have they changed after one month?

Which materials from your dustbin will rot away quickly on a dump? Which will rot away only slowly? Which will take a very long time? Which may never rot at all?

Keep
Britain
Tidy

2 Strong shapes

structure made from paper strips

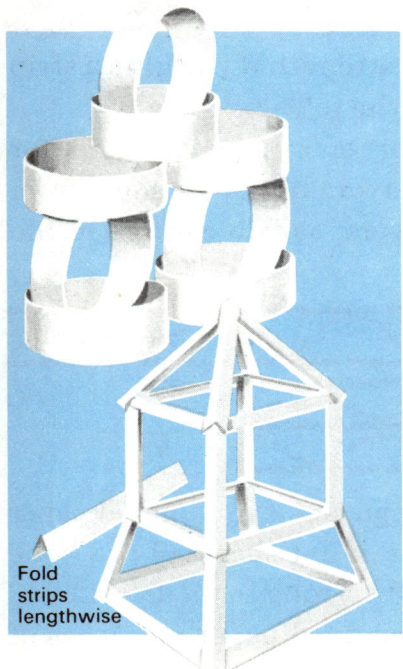

Fold strips lengthwise

structure made from *folded* paper strips

Building with strips of paper

1 Cut some paper into strips 3 or 4 cm wide. They may be all the same length or of several different lengths.

See what structures you can build with these strips without folding them. Use pins, paper-clips or glue for fixing. You will get some interesting shapes but nothing very tall or strong.

2 Next try folding all your strips of paper lengthwise before you start building.

See what kind of structure you can make now. How is it different from the structures built from unfolded paper?

How high can you make it?

Simply folding the pieces of paper has made them stronger. The strength of things depends on their shapes as well as what they are made of.

Folds, bends and girders

1 Cut four pieces of thin card or stiff paper all the same size, say 20 cm long and 12 cm wide.

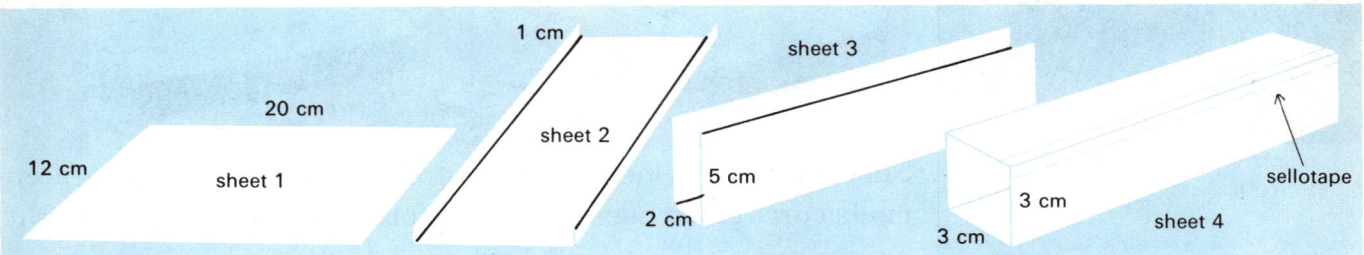

12 cm — sheet 1 — 20 cm

1 cm — sheet 2 — 2 cm

sheet 3 — 5 cm

sheet 4 — 3 cm — 3 cm — sellotape

Keep one piece flat but bend the others as shown in the pictures. Support each one at the ends. What weight will each carry before it bends or collapses? Coins or large washers are useful for weights.

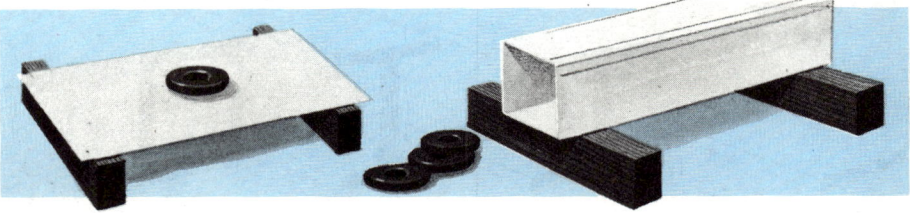

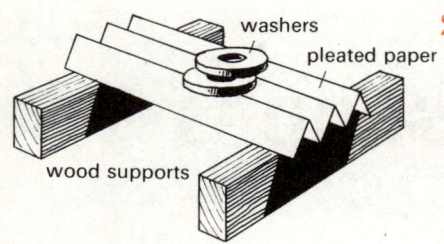

washers

pleated paper

wood supports

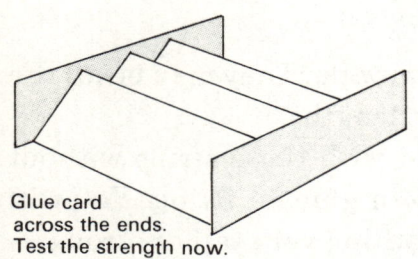

Glue card
across the ends.
Test the strength now.

2 Does pleating a piece of paper or card make it stronger? Try it. Try different sizes and shapes for the folds, like those in the next drawing.

What difference does it make to the strength if you glue a strip of paper or card across the ends of the folds?

3 Corrugated card is often used for making boxes and for packing. Some is made from two layers of paper, some from three layers and some from five layers, as you can see in the drawing.

Invent a simple experiment to see if pieces of corrugated paper are stronger than the same number of flat sheets of the same material. Is corrugated paper equally strong in all directions? Try supporting some pieces of corrugated paper on pieces of wood as shown in the drawings A and B.

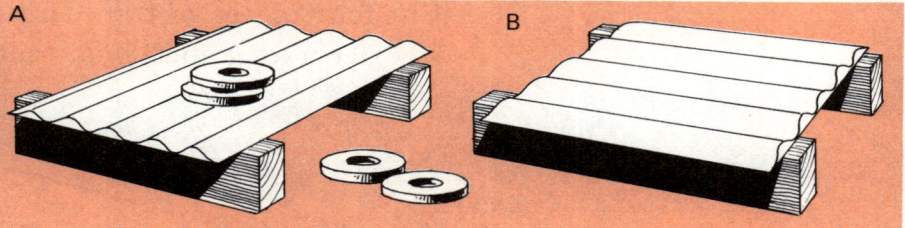

A
B

A carport with corrugated plastic roof

Sheets of iron, plastic or asbestos used for roofing are often made corrugated because in this form they are light in weight yet strong.

4 Look at the shapes of strong girders in buildings and bridges. You can see them best when the structures are being built.

Girder sections

box-girder bridge section

Make some model girders out of card. Make them in the different shapes. Test how strong the shapes are.

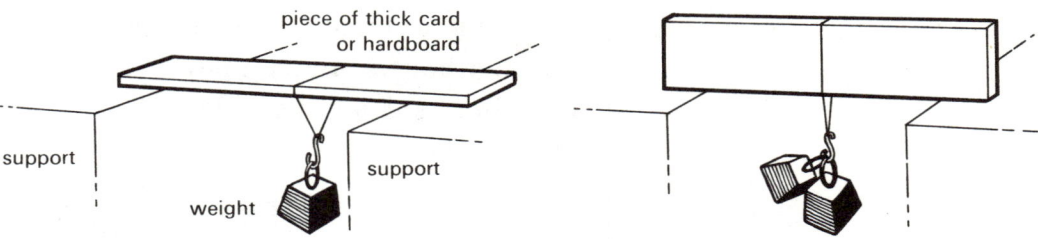

piece of thick card or hardboard

support

support

weight

5 Test some strips of wood flat and standing, as shown in the picture, to find out which way is the best for weight-carrying.

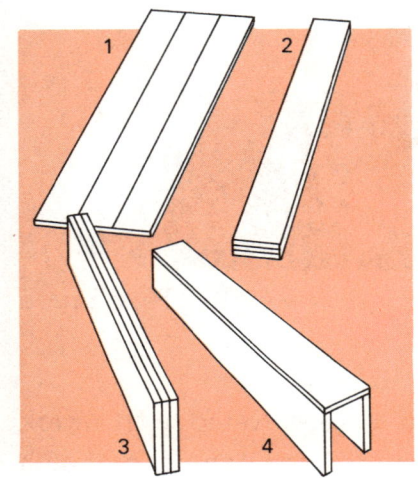

6 Suppose you had three planks each 2 metres long, 30 cm wide and 3 cm thick to make a bridge over a stream $1\frac{1}{2}$ metres wide, so that you could wheel over a loaded wheelbarrow. You have nails and a hammer to fix them together if you want to. Which would be the best way to use the planks to make the bridge strong and easy to walk on?

Four ideas are shown in the picture. How well would they work?

Is there a better way of using the planks?

Have you just guessed, or have you been a proper scientist and done some tests?

Arches

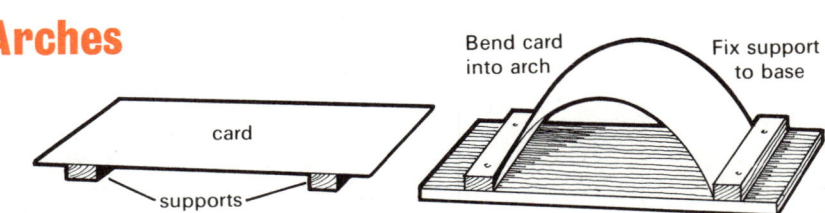

Bend card into arch

Fix support to base

card

supports

Put coins, washers or similar small weights on the centre of a piece of card arranged as shown and see how many are needed to make it sag.

Now bend the same card into an arch shape. Stop the ends from moving outwards. Is this shape any stronger? Put your small weights on top to find out. Compare some high arches with some flatter ones.

Look for arches and domes in the buildings you see around you and in pictures.

Building with newspaper

1 What can you build with sheets of newspaper? Have a go. How can you make a newspaper bridge between two tables or chairs about a metre apart? You might paste twenty sheets together and let them dry. What weight will twenty pasted sheets take?

How many sheets are needed to carry *your* weight? How much would this bridge itself weigh?

2 Next, roll some newspaper sheets into tubes.

Do this by rolling each sheet round a stick. Remove the stick and fasten each roll with sticky tape. Then tape together a number of the tubes to make a bundle.

Test the strength of the bundle. Is it stronger than the same number of flat sheets of paper pasted together?

sellotape

bundle of newspaper tubes

Newspaper tubes are good things to use for building other structures.

Can you use them to make a tower high enough to reach the ceiling? Can you make a dome or other shape big enough to sit or stand under? What other structures can you build?

Can you increase the strength of the tubes? One good way is to paste the sheets of newspaper with wallpaper paste and roll each round a smooth rod while the paste is still wet. If this is done not too tightly the tubes will slide off the rod and may be put on one side to dry. Test the dry pasted tubes to see how much stronger they are than un-pasted ones. You can simply try bending them in your hands to get the feel of their strength, but the experiment in the next picture on page 23 is better.

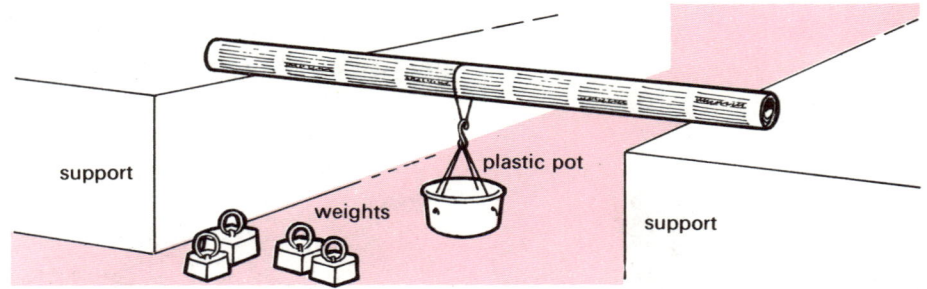

support   plastic pot

weights

support

Be sure to make it a fair test. Use tubes made from the same paper, see that they have the same diameter and that there is always the same length of tube between the supports.

The best size for strong tubes

How can tubes be made as strong as possible? Should they be wide or narrow? The only way to find out is to do some tests. To be fair, the same amount of material must be used for each tube, so the wide ones will have thinner walls and the narrow ones will have thicker walls. At a guess, it seems that the narrow tubes with thick walls might be the stronger. Suppose the material is used to make a solid rod instead of a tube. Is that stronger still? Try it out.

1 Get several sheets of paper all the same size. Roll each sheet into a tube with a different diameter. Hold down the edge of the paper with *small* bits of sticky tape as shown in the picture. The widest diameter tube will have just one thickness of paper. The smaller diameter tubes will have walls made of more thicknesses of paper but every tube will have the same amount of paper in it. Finally, roll a sheet of paper tightly so that it is like a solid rod.

Test the strength of the tubes as shown in the next drawing.

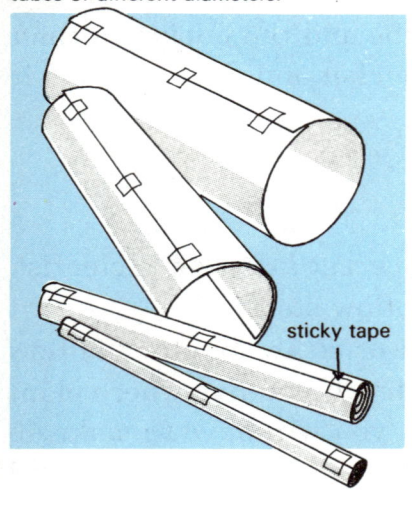

Roll the same sized sheets of paper into tubes of different diameters.

sticky tape

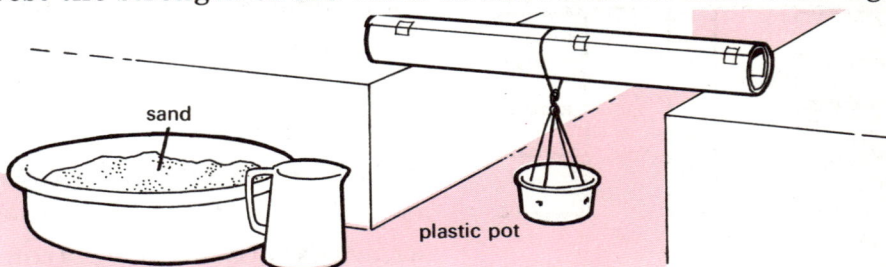

sand

plastic pot

Pour sand slowly into the pot until the tube buckles. Then weigh the sand and the pot. Make a record of your results. For *each* test you must have *exactly* the same length of tube between the supports and see that the weight hangs from the centre. From your results plot a graph like the one shown here.

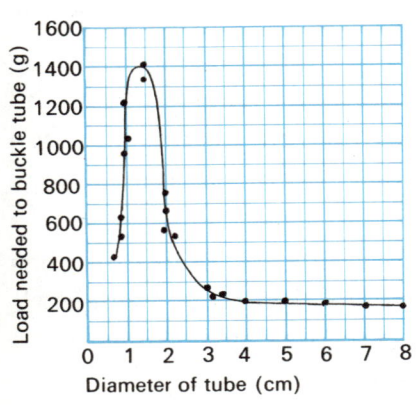

Load needed to buckle tube (g) / Diameter of tube (cm)

Is there a 'best diameter' for strength, or do the tubes simply get stronger as their diameter gets smaller?

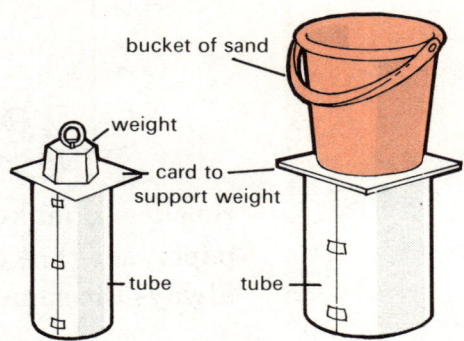

bucket of sand

weight

card to support weight

tube   tube

Pour sand into one side only

cardboard partition

2 When tubes are placed vertically, it is interesting to see how very strong they are.

The picture above shows you how to test some tubes in this position. Test some tubes yourself.

What happens if you do not load them evenly? You can try this most easily with the large tube and the bucket of sand. Put a cardboard partition in the bucket and pour the sand in one side only.

Bundles

Leonardo da Vinci (1452–1519) the great artist and scientist, experimented with rushes with hollow stems.

1 Get some rushes or some lengths of straw or other hollow plant stems. It is best to let them dry. If you live where plant stems are not very easy to collect, you will have to make do with drinking-straws, preferably the rolled paper kind and not the plastic ones.

Arrange two supports about 20–30 cm apart, and place one of your stems or straws across. Find what weight is needed to make it bend 5 mm in the middle.

Wire can be used to make the small hooks, and washers make good weights.

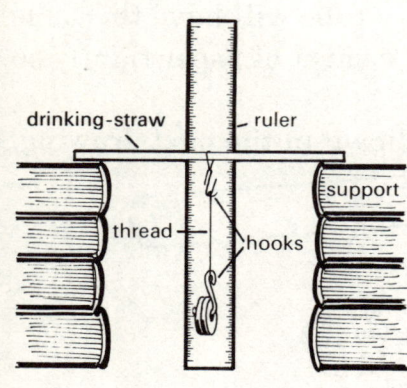

drinking-straw — ruler

thread

hooks

support

Next try *two* stems or straws held together with a rubber band at each end. Then try three, then four and so on. Are two straws exactly twice as strong as one? Are three straws three times as strong as one?

Leonardo said not. He said that when he bound rushes together, each rush supported twelve times the weight it could support alone. Do you get a result like this? Does it not depend on *how many* you bind together?

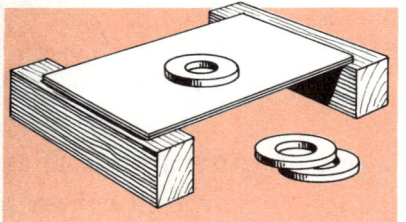

How strong are two flat sheets glued together?

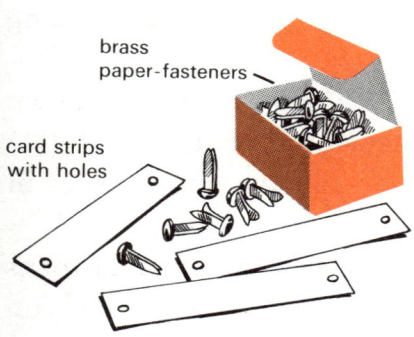

brass paper-fasteners

card strips with holes

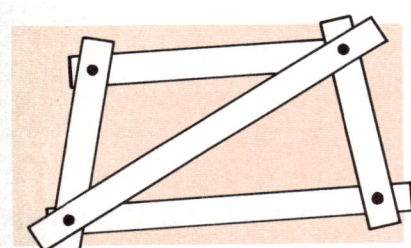

Instead of just binding the ends, try binding the bundle along its whole length. Does this make a difference to the strength? What happens if you *glue* the stems, or straws, firmly together?

2 Do these experiments work for materials in shapes different from tubes? For example, try flat pieces of stiff paper or thin card.

Are two pieces glued together just twice as strong as one piece? Thin sheets of material glued together like this are called *laminations*.

Rigid frameworks

1 If you make some frameworks of strips, like those in the picture, and fix the corners with brass paper-fasteners (or pins or nails) you will find that they are not rigid. They flop about. These frames would not do for building anything.

You might glue the corners or use *two* fasteners at each corner, but the structure would still not be very strong. The strength would depend on the strength of the joints. But if you put a strip across, as in the picture, then you will find that the frame is rigid even with single, loosely-fitting paper-fasteners at the corners.

Make some frameworks with three, four, five, six, etc. sides.

Which of these shapes is rigid to start with? Try different ways of dividing up the others with strips to make those rigid too. Generally, you will find that you need to form triangles. The strengthening strips need not always go to the corners. Try these:

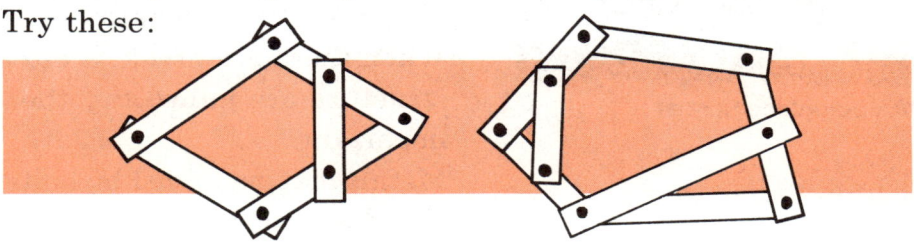

Steel framework of building

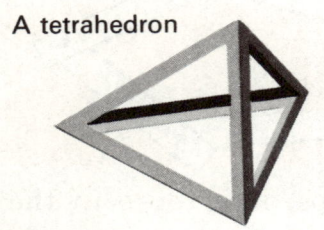
A tetrahedron

Tower cranes

An aeroplane hangar

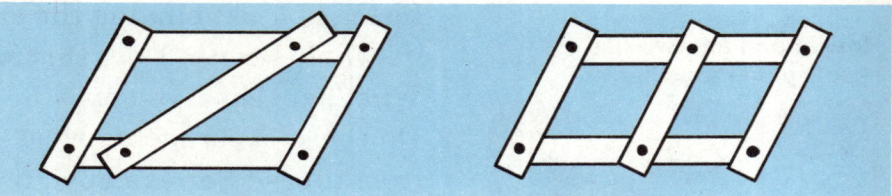

Make the frameworks shown in the two pictures above. They are very much alike but what difference do you find between them?

Look for triangle shapes used in cranes, bridges, scaffolding and other parts of buildings.

2 An even stronger framework is a *tetrahedron*. This is shown in the picture. It is like a three-dimensional triangle.

Can you see this shape in the cranes and the hangar roof in the pictures?

You don't *always* have to form triangles to make frameworks rigid. Try the frameworks in the next drawings and see if you can find some more of your own.

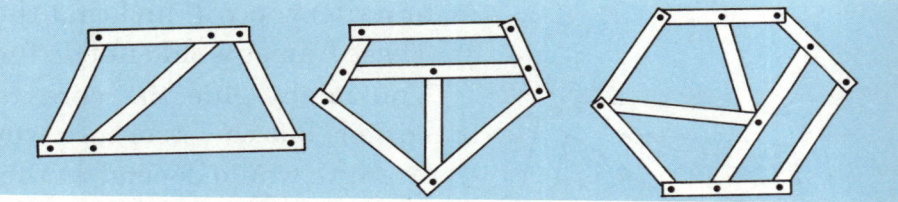

3 You can make a good model bridge from thin strips of balsa wood. (3 mm square section is a good size.) Stick the strips with balsa cement. Is the bridge strong enough to carry the weight of a brick?

Joints

Very often the strength of structures depends on their joints. When there is a break, it often takes place where parts of the structure are joined together. Fixing joints strongly is very important.

Metals may be joined by bolts, rivets or welding.

Wood is joined by nails, screws or glue.

Glues

If you look around shops you will see that there are lots of different kinds of glues. Some are for sticking paper together, some for wallpaper, some for wood, some for plastics and some for pots. Some even claim they will stick anything to anything! See how many different types of glue your class can collect.

Try out each glue, doing exactly what it says in the instructions. Find out what materials the glues *will* stick together. For example, will any of them stick metals together, a piece of rubber to a piece of plastic, or wood to concrete? You will have many more ideas like these to try. See whether the claims on the packets are true. **Never get any glue on your hands.** Some glues will stick fingers very firmly together!

Which glue is the strongest?

glued edge

glued edge

Pick out all the glues which will stick wood to wood, and then find which is the strongest. You will need some pieces of wood all the same size (e.g. 10 cm × 3 cm × 4 mm). Glue two ends together as shown in the drawing. Do this with each glue that you picked out, and don't forget to number or label the pieces of wood or you will not be able to remember which is which.

Be quite fair and follow the instructions correctly for each glue. Rest the glued pieces on something flat and give the glue plenty of time to set firmly.

Now you can test the strength of the join as shown in the next picture. Hold one end down firmly and hang a plastic pot on the other end. Put marbles or other small weights in the pot, one at a time, until the joint breaks. Be sure to have a box or tray immediately under the pot so that the marbles don't scatter when this happens.

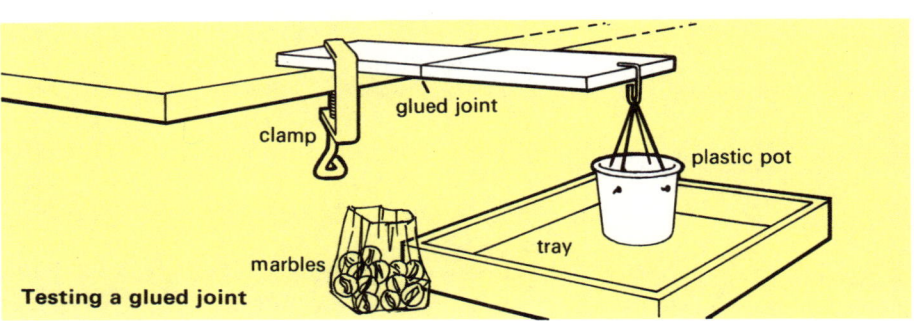

Testing a glued joint

clamp

glued joint

plastic pot

marbles

tray

You will probably be quite surprised at the strength of some glues.

Sticky tapes

Which is the stickiest sticky tape? First you can collect together plastic tapes, insulating tapes, sticking-plaster, masking tapes, etc., of different makes.

1 Take some pieces the same size. Press each firmly to a table-top with a rather long elastic band going underneath, as shown in the drawing. (You may have to experiment a little to find just the *right* elastic band.)

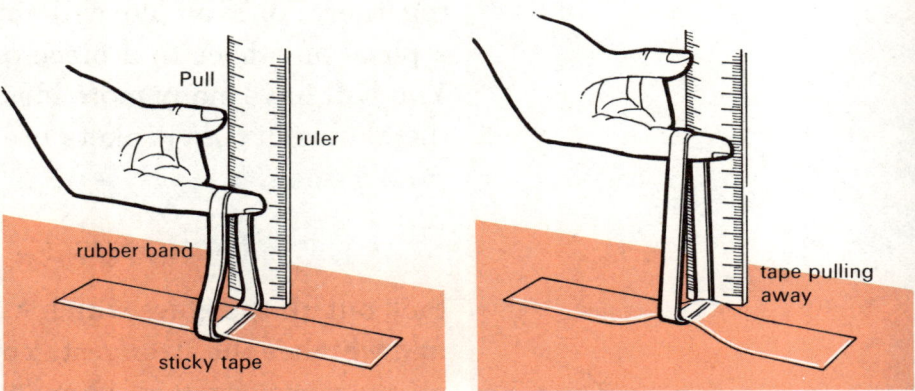

Measure how far the elastic band has stretched when the sticky strip begins to pull away from the table. Write down the *longest* stretch noticed. Will one test be enough?

Next, using the same elastic band, test the other sticky strips in turn. Write down your results. Which is the stickiest?

2 Now do some tests to find what difference the *width* of the tape makes to its sticking power. Do the test again, using the same type of tape each time, but in different widths.

3 Springy things

Plastic and elastic

In Chapter 2, we talk about rigid structures, but we do not always want the things we construct to be rigid. For example, we need springy arrangements for car suspensions, mattresses, cushions, diving-boards, balls, tyres, trampolines and the springs which drive clocks and toys. The materials we need for these things have to be *elastic*. Rubber, some metals, some plastics and the air are *elastic*. This means that if they are pushed or pulled out of shape, they spring back again. Some materials are not elastic but are *plastic*.

Clay, putty and plasticine are *plastic*. This means that when you push or pull them out of shape, they stay in their new shapes and do not spring back again.

> Be careful about this word, *plastic*. We use many things which are made of the materials called plastics, but hardly any of them are *plastic* like clay.

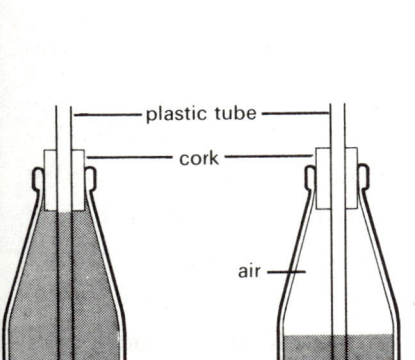

1 You can feel how elastic the air is by pulling out the handle of a bicycle-pump, putting your finger tightly over the open end, and then trying to push the handle in again.

2 There is also a 'trick' experiment which shows that air is elastic but water is not. The picture shows you how to do it. Notice that the tube goes well under the water, almost to the bottom of the bottle. See that the cork is a good fit.

Try to blow down the tube when the bottle is full of water. Can you force any air in? Then try again with the bottle *half* full of water. How does the result tell you that air is springy?

plastic tube
cork
air
water

Springs and rubber bands

Collect together several sizes of rubber bands, pieces of elastic, and springs of the kind which stretch. These springs are not always easy to find, but your teacher should be able to get some. If you have an old 'Slinky' toy, it is easy to cut this up and

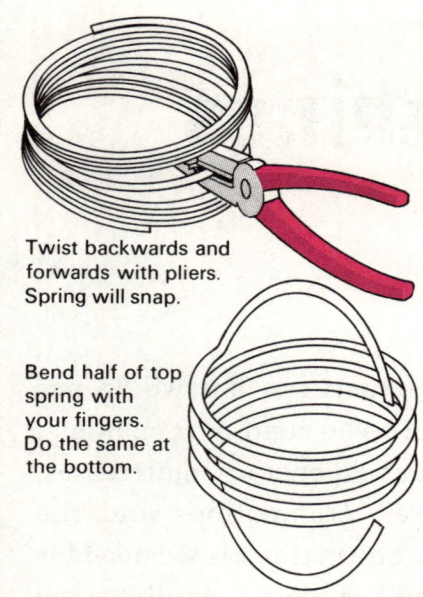

Twist backwards and forwards with pliers. Spring will snap.

Bend half of top spring with your fingers. Do the same at the bottom.

make some first-rate springs for the experiments. The picture shows how to do it.

Make some 5, 10, 15 and 20-turn springs in this way. Be gentle with them. If you pull them too hard they will not return to their proper shape again.

Stretching

1 Hang a spring by one of its coils as shown in the picture below. Glue a matchstick or card pointer to the lower end.

Use marbles, washers, nails, bolts or nuts as weights to put in the plastic pot.

Glue pointer to spring

2 kinds of spring for stretching

Measure and record how much the spring stretches for different weights. Make a graph from your results. The picture shows the start of one. What do you find?

Now repeat the experiment with a spring with a different number of turns.

Make a graph for this one on the same paper as the first. How is it different?

Try more springs, each with a different number of turns.

Next try the experiment with a rubber band instead of a spring. Are the results similar? Try several different rubber bands.

2 For the next experiment, find out what happens when two similar rubber bands are arranged end to end, as in drawing A. See how much they stretch with one marble in the pot, then with two, then three and so on.

Do the two bands together stretch more or less than one band by itself?

Now arrange the same two bands as shown in picture B. How do they stretch now? Do they stretch the same as when they were fastened end to end?

Do they stretch the same as a single band?

Draw graphs to show what you find out.

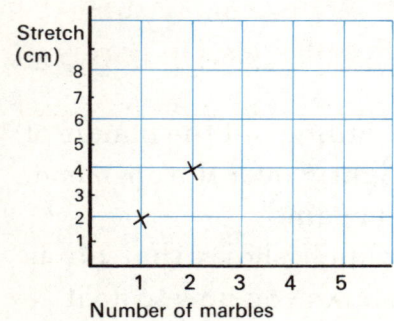

Stretch (cm)

8
7
6
5
4
3
2
1

1   2   3   4   5
Number of marbles

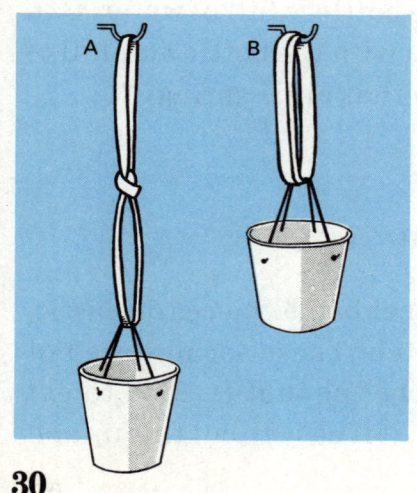

A   B

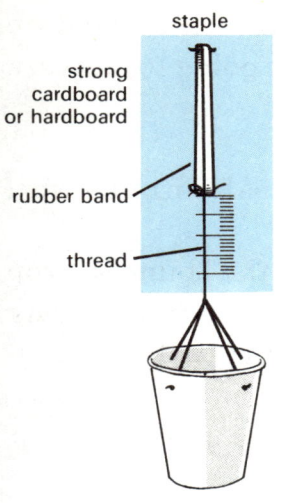

staple

strong cardboard or hardboard

rubber band

thread

Weighing machines

1 The picture shows how to make an elastic-band weighing machine. If you put marbles in the pot one at a time, you can mark the scale in 'marbles-weight'. If you use washers, it will be in 'washers-weight'. If you have some proper metric weights you can make the marks show grams-weight.

2 If you have a compression spring, like the ones used in some mattresses, a good weighing machine can be made as shown in the next picture. The scale can be made by adding marbles, washers or weights the same as with the elastic-band weighing machine.

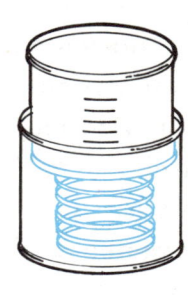

You need two similar shaped cans. One should fit loosely into the other. Put the spring into the bottom of the large can and fit the smaller can on top of the spring.

Bouncing

1 Will a heavy weight on the end of a spring or a rubber band bounce more quickly or more slowly than a light weight? Find out.

2 Try bouncing a weight on a single rubber band, then the same weight on two bands looped together in line, then on three bands looped together.

How does the bouncing-time change?

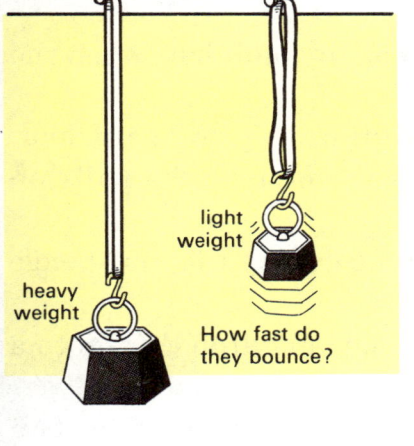

light weight

heavy weight

How fast do they bounce?

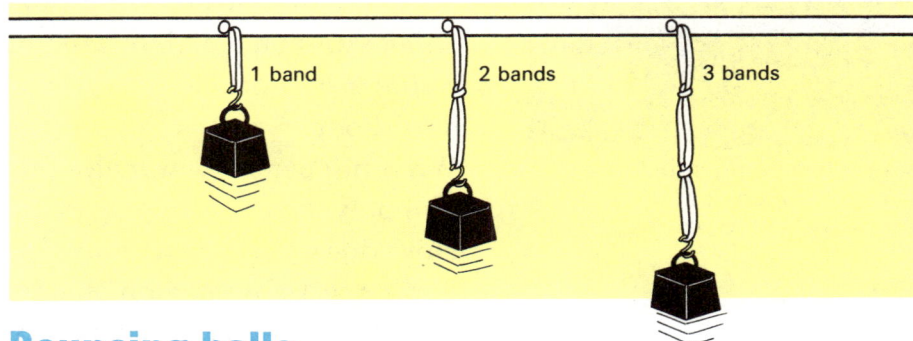

1 band

2 bands

3 bands

Bouncing balls

See how many different kinds of ball you can collect. There are footballs, golf balls, tennis-balls, cricket balls, table tennis balls, rubber balls, plastic balls, glass marbles, steel ball-bearings and many others.

Write down what each ball is made of. Which kind of ball bounces best?

1 How can you test bouncing? You must be quite fair and drop each ball from the same height. How are you going to measure the height of the bounce?

Will one test be enough?

When you have finished testing, arrange the balls in order with the best bouncer first in the line.

2 Instead of trying to measure the heights of the bounces, drop each ball from the same height as before and count how many times it bounces before it stops. Again, arrange the balls in order with the best bouncer first.

Does this method give the same result as the first test? Is it easier to do? Is it as accurate as the first?

3 *How* does a ball bounce?

Powder some chalk and put a small patch of the powder on the floor. Drop a dark-coloured rubber ball on the powder patch. Catch the ball after the first bounce and look at it. How big is the white patch on the ball? What does the white patch tell you happened to the ball as it hit the floor?

Next try the same experiment with a hard cricket ball. Is the white patch on the cricket ball the same as that on the rubber ball? Try a football.

The photograph shows what happens to a golf ball when the club hits it.

4 You probably did all the bouncing tests on the same hard floor. Would it make any difference if you bounced the balls on different surfaces?

Try them on carpet, on a polystyrene ceiling tile, on sponge rubber and on a dish of sand or soft soil.

5 Make a ball of plasticine or damp clay and drop this on to a hard floor.

What happens? Now make the ball round again and drop it on to a polystyrene tile placed on the floor. What do you notice? How does your best bouncing ball which you found in the earlier tests bounce on the polystyrene tile? How well a ball bounces depends on what it bounces on as well as what it is made of.

A golf ball being struck with a club. Look at the shape of the ball!

4 Hot and cold

Thermometers

Clinical thermometers

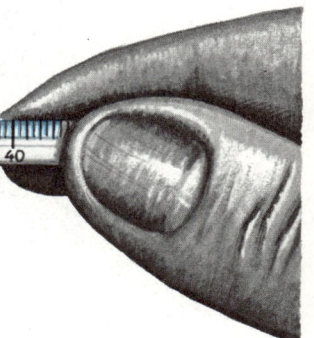

A clinical thermometer

Have you ever been ill and had your *temperature* taken with a thermometer like the one in the picture? The bulb is placed under your tongue. The silvery liquid inside (mercury) pushes up through the narrow part of the tube to stop at a certain height up the scale. This shows how hot you are.

The divisions on the scale show *degrees,* and the ones we normally use are called degrees Celsius. (Celsius was the name of the man who invented this scale.) We write 'degrees Celsius', '°C' for short. Normally your temperature is about 36·9°C but when you are ill it can be higher than this.

Once the mercury has pushed past the narrow section of the tube it will not go back into the bulb until it is shaken down. What is the advantage of this?

Room thermometers

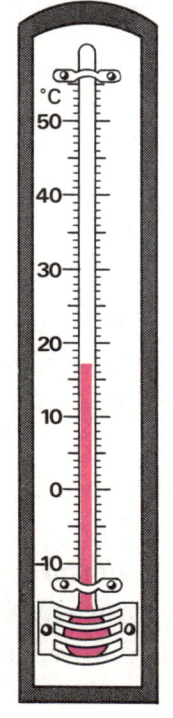

A room thermometer

Another thermometer you will have seen is a room thermometer. The liquid in a room thermometer is *alcohol*, not mercury, and it is often coloured red or blue.

1 Get a room thermometer and measure temperatures in many different places. Try outside in the sunshine and shade, under the cold-water tap, over a radiator and many other places. Do not put a room thermometer in *very* hot water or in an oven, because room thermometers are only made to read up to about 55°C. If you put them in places hotter than this, the liquid inside will force off the top of the glass tube. Special thermometers are made to read higher temperatures.

WARNING

In the experiments in this chapter where you have to use hot water, *BE CAREFUL*. Use hot water but *NOT* near to boiling which will scald.

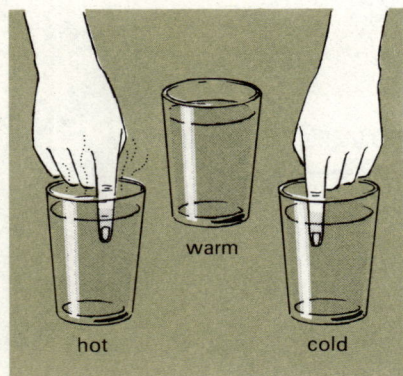

warm

hot   cold

warm

hot   cold

2 What is the coldest temperature you can make?

Put a thermometer in some ice from a refrigerator. What is the temperature?

Does it get colder if you add more ice?

Now mix the melting ice with various substances. First try a teaspoonful of salt.

Then, starting with fresh ice each time, try a teaspoonful each of sugar, vinegar, milk and sand.

Do the experiments again with two teaspoonfuls, then three. Write down the lowest temperature you get in each experiment.

3 Find out what the temperature is in your refrigerator. Is it the same in all parts? What is the temperature in a freezer?

Feeling temperatures

Our bodies can judge some temperatures roughly. We can feel whether it is a warm day or a cold day and whether the bath water is just right, but we cannot test temperatures accurately without a thermometer.

Have three tumblers of water as shown in the picture. The hot water should be as hot as you can bear your finger in.

Put one finger in the cold water tumbler and another in the hot water tumbler. Leave them there for a minute. Then move both fingers into the warm water tumbler.

One finger will feel that the warm water is hot and the other finger will feel that it is cold.

Expansion and contraction

Liquids

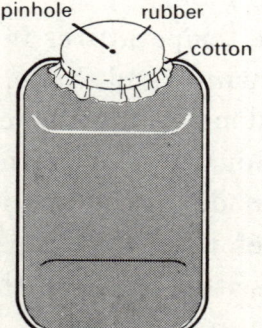

pinhole   rubber

cotton

1 Fill a small glass bottle right up to the top with cold water. Cut a piece of rubber from a balloon, stretch it over the top of the bottle and tie it down with cotton. Make a small hole with a pin in the middle of the rubber. There must be no air in the bottle.

Now put the bottle up to its neck into a basin of hot water. What happens? Can you explain it?

The same thing happens to the liquid in a thermometer. As it becomes warmer it rises up the tube because it expands. When the liquid cools it contracts and goes down the tube again.

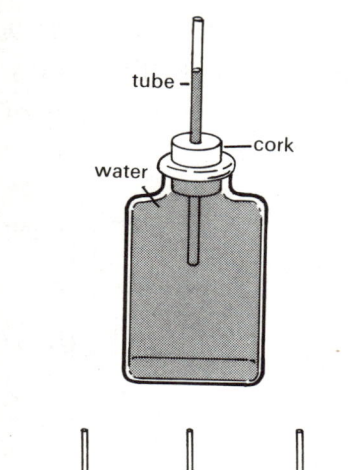

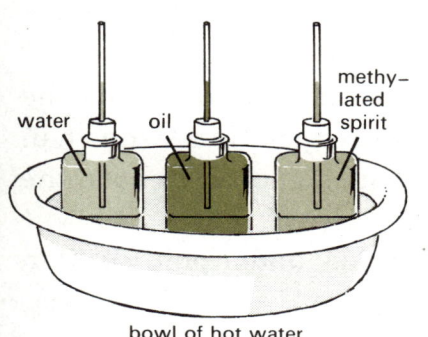

bowl of hot water

2 Fill up a bottle with water and then fit a cork and tube as shown.

As you push in the cork, some water will rise up the tube. Now warm the bottle of water over a radiator, or simply with your hands. Watch the water rise in the tube as it expands. Cool the bottle. Watch the column of water go down again.

3 Now fit two more bottles with tubes exactly like the first one. Fill one with cooking oil or cycle oil and one with methylated spirit. (BE CAREFUL! KEEP IT AWAY FROM A FLAME.) By pushing in the corks or loosening them, get the liquids in all three tubes to the same level. Put the three bottles into a bowl of hot water at the same time. What do you see? Do all the liquids expand equally? Which liquid expands quickly? Which is sluggish? Let the bottles cool again. Does the liquid which expands most quickly also contract most quickly?

Solids

Solid materials also expand and contract as they are warmed and cooled, but not nearly as much as liquids do. Have you seen the small gaps left between the sections of motorway bridges so that they can expand a little on hot summer days?

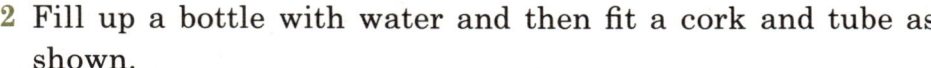

Expansion gap on motorway bridge

1 Fix about 3 metres of bare copper wire tightly between two firm supports. From the middle hang a weight of some kind, so that it is just a small distance from the floor.

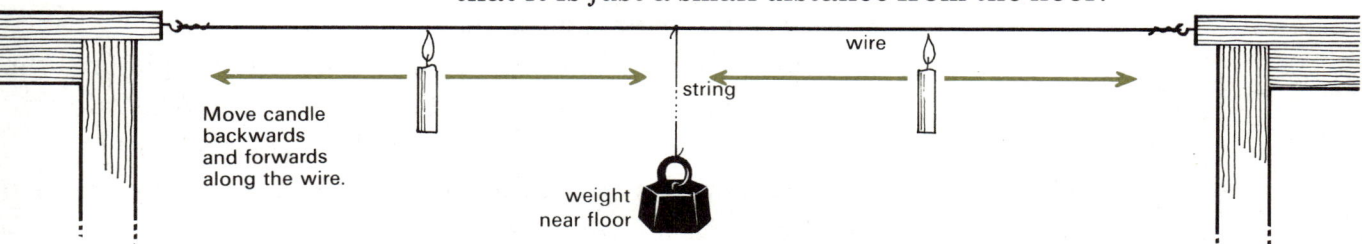

Now warm the wire. The easiest way to do this is to have two people, each one heating half the wire by moving a lighted candle slowly backwards and forwards along it. How can you tell that the copper expands? Does it contract when it cools?

A lot of experiments would have to be done to find out whether *all* materials expand when they are heated. Most of them do, but perhaps you will be interested in the strange case of rubber. It actually contracts when it is heated.

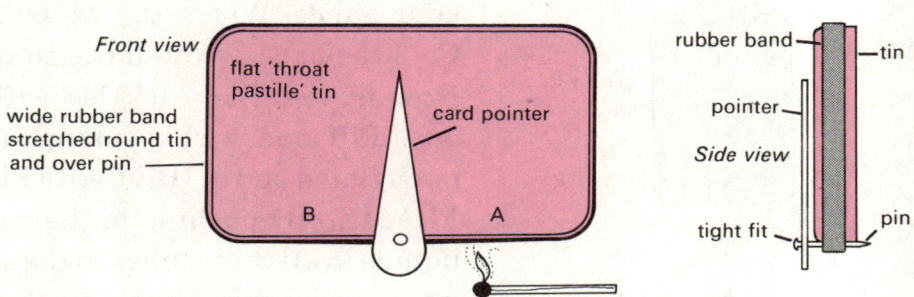

Front view

wide rubber band stretched round tin and over pin

flat 'throat pastille' tin

card pointer

B   A

rubber band — tin

pointer

Side view

tight fit — pin

2   You can see this by making the apparatus shown. Hold the flame of a match under the rubber band in position A shown in the picture. Heat the band (but NOT long enough to burn the rubber). What does the pointer do?

Now hold the flame under the other side, under point B. Can you explain what happens?

Gases

1   Take the cap off an empty washing-up liquid bottle and squeeze in the sides. Hold it like this while you put the cap on again. Warm the flattened bottle over a radiator, then cool it under a cold-water tap.

How do you explain what happens?

2   Fit a small bottle with a cork and tube. First cool the bottle in cold water and then hold it as shown in the drawing, with the end of the tube under water.

Now warm the bottle by placing your hands close around it. What do you see? Can you explain what happens?

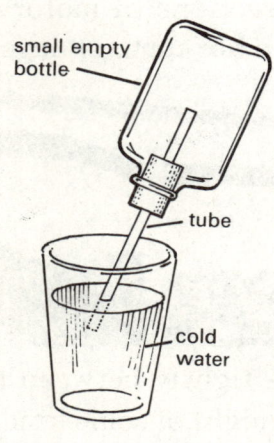

small empty bottle

tube

cold water

3   For this experiment, use a balloon which has been well stretched by having been blown up several times.

Cool a large 'pop' bottle in cold water or in a refrigerator.

Fit the balloon over the top of the bottle.

Now warm the bottle (and the air inside it) by placing the bottle in a bucket of hot (not boiling) water. What do you expect to happen? Does it? — as much as you thought?

These experiments have all been with air, but other gases also expand quite a lot when they are heated.

Because air expands like this, hot air is lighter than the same volume of cold air. This causes hot air to rise through cooler air. Look again at Chapter 1 of Book 2 for some more experiments and a picture of a hot-air balloon.

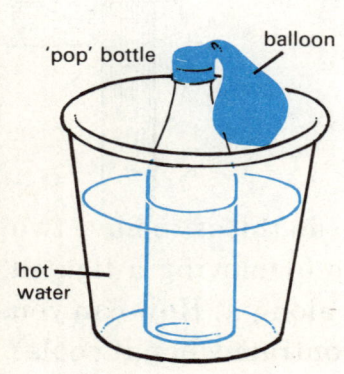

'pop' bottle

balloon

hot water

Hot water rises

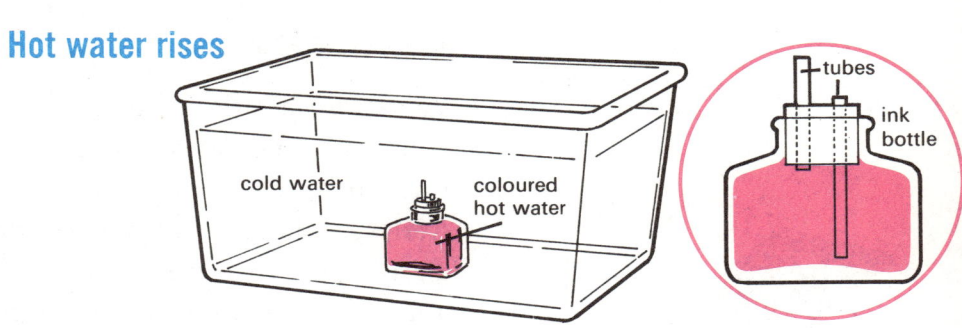

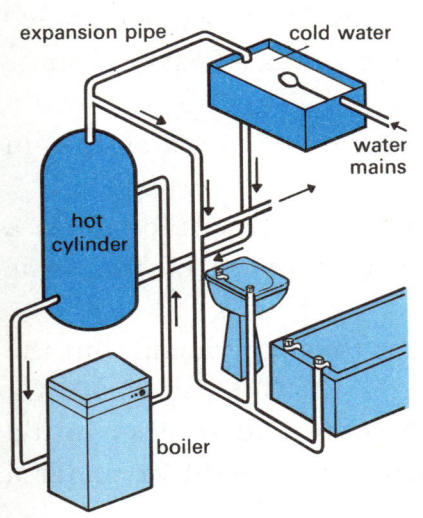

1　Hot water is also lighter than the same volume of cold water. Fill a small bottle completely with hot water which has been coloured with ink. Fit a cork and two tubes like those in the picture. When the cork is pushed in, there should not be any air bubbles under it.

Put the bottle at the bottom of an aquarium tank or a bucket full of cold water.

What happens? How long does the movement go on?

When you feel the hot-water cylinder in your house, which is the hottest part? Why?

2　Look at the diagram of a house hot-water system.

Remembering that hot water rises, work out how the water will move around when the boiler heats some of it.

What happens in a hot-water system when some water is drawn off from a hot tap?

Heat from the sun

How many ways of heating a house do you know of? What fuels are used? All the ways of heating houses are expensive because fuels are expensive. Perhaps we could use the heat coming directly from the sun more than we do at present. Many experiments are now going on to find out the best way of using the sun's heat. The pictures show a 'sun-furnace' which has been built in France and a house designed to use the heat from the sun to heat its hot-water supply.

Solar heating panels fitted in the roof of a house

A 'sun-furnace'

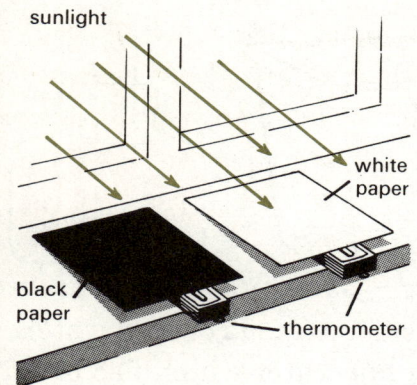

sunlight

white paper

black paper

thermometer

Absorbing heat rays

1 Take two similar thermometers and put one under a black sheet of paper and the other under a white sheet of paper. Put both in sunshine (or in the light from a table-lamp). Arrange them so that the light falls equally on both. After about half an hour, look at the thermometers. Is there any difference between the temperatures under the two papers?

2 Change the black paper for a red one and use a fresh piece of white paper. Do the experiment again. What happens now? Test other colours besides red. Which colour of paper absorbs heat rays the best?

Why do we prefer to wear white or light-coloured clothes in summer and in hot countries?

3 Now find out whether a paper with a dull surface gives a different result from a shiny-surfaced paper of the same colour.

4 You can use room thermometers for this experiment, but the all-glass laboratory type are better (0°–50°C or 0°–100°C). Put one thermometer in a dish of water and another into a similar dish of soil. Put both dishes in sunlight or under a lamp as shown in the picture.

water

soil

Record the temperatures of the soil and the water every ten minutes and make a graph from your results. Do soil and water heat up in the same way?

On a hot summer's day at the seaside, when the sand feels very warm, what do you notice about the temperature of the sea? Why is this?

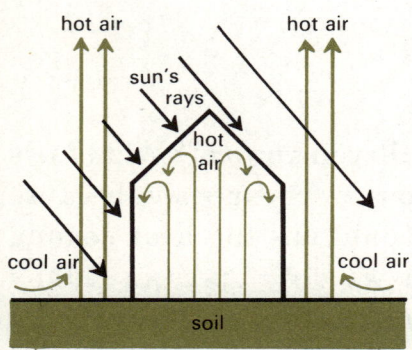

hot air

hot air

sun's rays

hot air

cool air

cool air

soil

How sunlight heats a greenhouse

Greenhouses

Gardeners make good use of the sun's heat in a greenhouse. The sun's rays come through the glass and heat up the soil, just as they do outside. The soil warms the air about it and the warm air rises. In the open, rising warm air carries much of the heat away from the soil but in the greenhouse the warm air cannot escape.

1 Try the experiment with a glass jar, a dish of soil and two thermometers as shown in the picture. If the sun is not shining use a desk-lamp as a substitute.

What do you notice about the readings of the two thermometers after an hour?

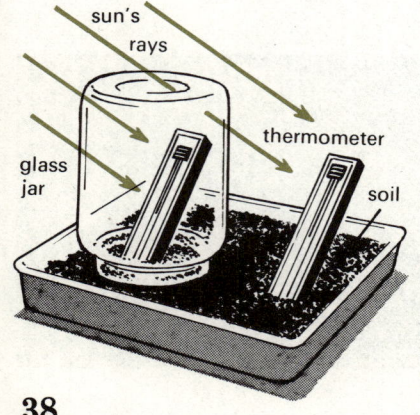

sun's rays

thermometer

glass jar

soil

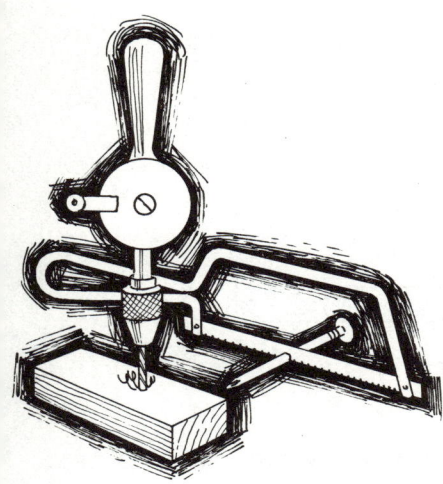

Spacecraft re-entering the earth's atmosphere

red-hot heat shield

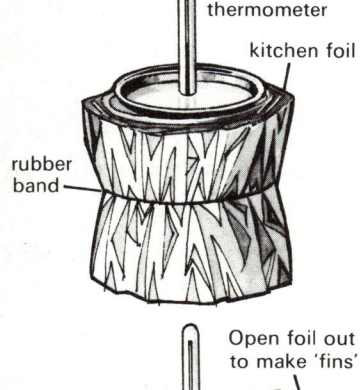

0–110°C thermometer

kitchen foil

rubber band

Open foil out to make 'fins'

2 Find out if a clear plastic bottle, or a plastic bag, works as well as the glass jar. Does a cloudy-white (translucent) plastic bottle work? Would a plastic greenhouse have any advantages over a glass one?

Friction produces heat

Use a hack-saw to saw through a large nail. Use a drill to make a hole in metal, or in a piece of hardwood. Feel the saw blade, the drill-bit, the nail and the metal or wood immediately after you have been working. Take care, they are sometimes hot enough to burn your skin!

Wherever there is friction, heat is produced. We rub our hands together to warm them in cold weather; a match head rubbed on the side of a matchbox produces enough heat to make it burst into flames, and the brake-blocks and wheel rims of your bicycle become quite hot when you have been using the brakes. What other examples of friction producing heat do you know?

The most spectacular one of all is shown in the picture. When a spacecraft re-enters the atmosphere, it is going so fast that the friction of the air makes it very hot indeed. To save the astronauts inside, the craft has a very special heat shield on the front which burns away, using up the heat and keeping the inside of the capsule reasonably cool.

Cooling

Look for the large cooling towers at electricity power-stations. Steam from the turbines becomes water again on the cool inside walls of these towers.

Take a can with a lid and make a hole for a thermometer as shown. Cut a strip of kitchen foil long enough to go around the can three or four times. Wrap this tightly round and hold it in place with a rubber band. Fill the can with hot water, record the temperature, and note the difference in the temperature after half an hour.

Now open out the foil and make it form 'fins' round the can as shown. Fill the can with hot water at the same temperature as before. How much does the temperature fall in half an hour? How does this compare with the first experiment? The bigger the surface area an object has, the more quickly it will cool.

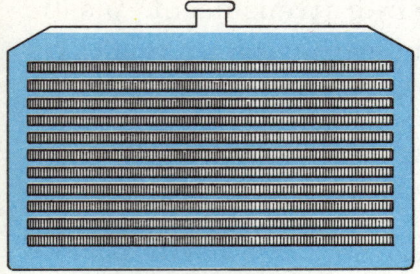

In some radiators the water tubes are horizontal —

in others they are vertical.

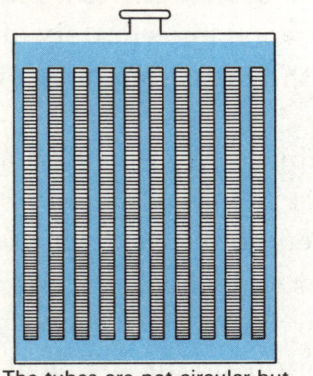

The tubes are not circular but have a cross-section like this ⊏⊐

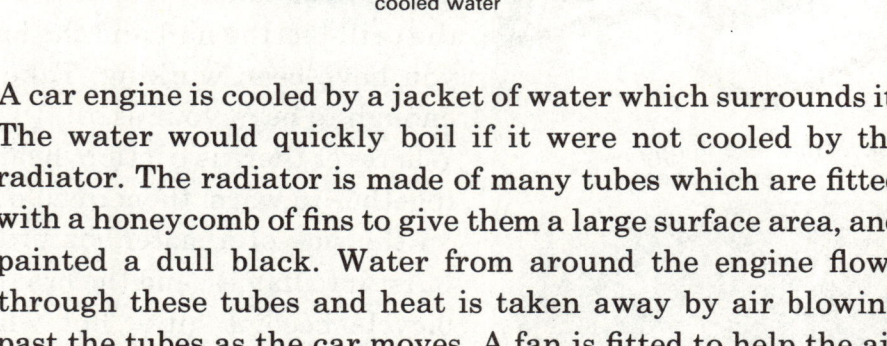

A car engine is cooled by a jacket of water which surrounds it. The water would quickly boil if it were not cooled by the radiator. The radiator is made of many tubes which are fitted with a honeycomb of fins to give them a large surface area, and painted a dull black. Water from around the engine flows through these tubes and heat is taken away by air blowing past the tubes as the car moves. A fan is fitted to help the air flow through more quickly.

Do large objects cool more quickly than small ones?

Get several cans, with lids, all of the same shape but of different sizes. Fit them with thermometers (0°–110°C) as shown in the picture. Fill each can with hot water at about the same temperature. Write down the temperature in each can every ten minutes, and make a cooling graph for each can.

(There will be 3 graph-lines on your paper if you use 3 cans. Only one is shown in the drawing here, so that you are not told the answer.)

What do you find out?

What does this tell you about large and small animals in winter?

Why do babies need to be wrapped up in plenty of woolly clothes, many more than we ourselves need, in cold weather?

Evaporation causes cooling

1 Pour a little methylated spirit into the palm of your hand. It will evaporate. The heat needed for evaporation comes from the heat in your hand.

2 To keep milk fresh in hot weather, people sometimes cover the bottle with a cloth soaked in water. Does it work? Try this experiment.

0–110°C thermometers

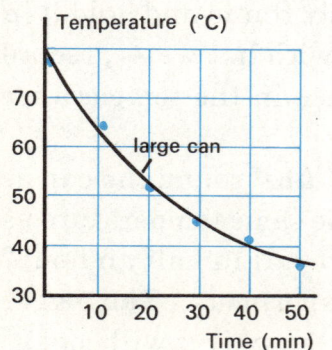

cans of same shape but different sizes

40

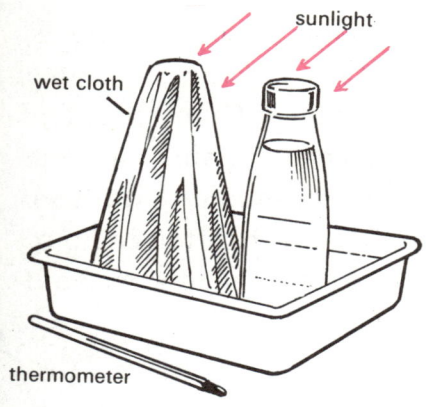

wet cloth
sunlight
thermometer

Stand two bottles full of water in a shallow dish of water. Put them in the sunshine. Cover one of the bottles with a cloth and see that the cloth is kept wet. Measure the temperature of the water in the bottles before you start and again after an hour or two. What do you notice about the temperatures of the two bottles? Has the evaporation of water from the cloth caused cooling?

When we work hard or play active games we sweat. Beads of perspiration are given out from small holes in our skin called pores. As this evaporates into the air we are cooled a little. Dogs have no pores for sweating in the skin covering their bodies. A dog's tongue is the only place from which it can perspire. That is why you frequently see dogs with their tongues hanging out, like the one in the picture.

Heat insulators

When we have heated something we often want to keep it hot. The pictures show a wrapped-up hot-water cylinder, and a man putting thick strips of fibre-glass in the roof-space of his house to keep the warmth in the rooms below.

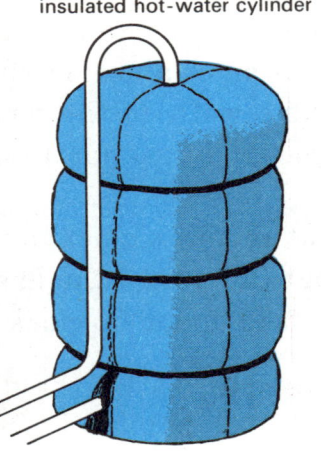

insulated hot-water cylinder

Insulating the loft of a house helps to keep the heat in

Materials which do not allow heat to pass through them easily are called heat *insulators*. Try these experiments.

1 Set up a can of hot water as shown in the drawing.
Take the temperature of the water every 30 minutes and record it on a graph.
Do the experiment with other packing materials around the can. You could try hay, wood shavings, cotton wool, etc.
Compare your results from each experiment. Which of the materials you used is the best insulator?

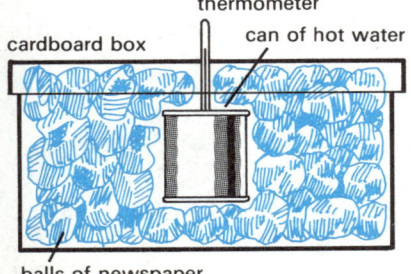

cardboard box
thermometer
can of hot water
balls of newspaper

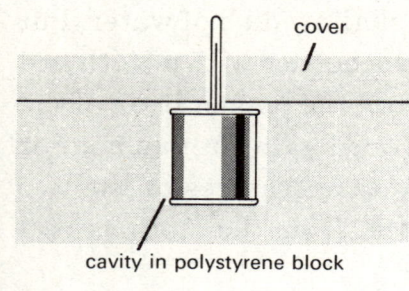

cover

cavity in polystyrene block

2 Test how good a heat insulator expanded polystyrene is. (This material is often used for packing.)

Use the can from the previous experiment. Get a block of polystyrene and make a hole in it to fit the can. Make a cover of polystyrene, too. Fill the can with hot water and write down the temperature of the water. Put on the cover. After an hour, take the temperature again. Now do the experiment without the cover. What do you notice about the results?

Air is a good heat insulator

If air is held still around a hot thing it will keep heat in. If it is free to move, it will rise when it becomes warm, carrying heat away and allowing cooler air to move into its place. The materials we used in the experiment on heat insulators trapped air around the can. The polystyrene had large amounts of air trapped *inside* it when it was made.

Birds use air as a heat insulator in winter. Have you seen a bird with its feathers fluffed out like the robin in the picture? By fluffing its feathers, the bird holds still a layer of air around its body. The layer of air keeps the warmth in. Our clothes trap air around our own bodies like this.

By fluffing out its feathers the robin holds a layer of air around its body

A vacuum flask

The next picture shows a vacuum flask. A vacuum (empty space) is a very good insulator indeed. As much air as possible is pumped out from between the double walls of the flask. Drinks will stay hot or cold inside a vacuum flask for a long time.

Put some hot water in a vacuum flask at about 9 o'clock in the morning, take the temperature and put the top on the flask. See what the temperature of the water is at about 4 o'clock in the afternoon.

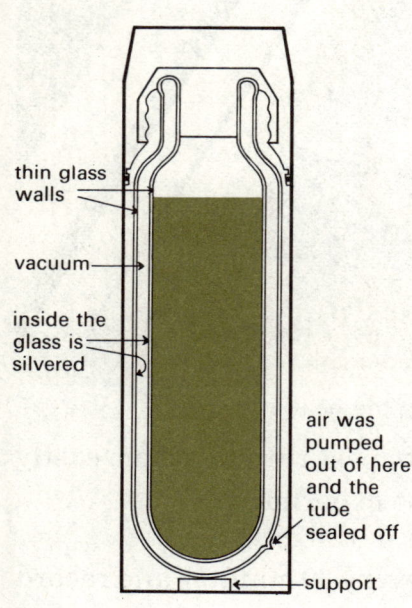

thin glass walls

vacuum

inside the glass is silvered

air was pumped out of here and the tube sealed off

support

Another use for heat insulators

We use heat insulators for handles of kettles, irons and pans and for making oven gloves and table mats, so that we do not burn our hands or mark the furniture. Find out what materials are used for making these things.

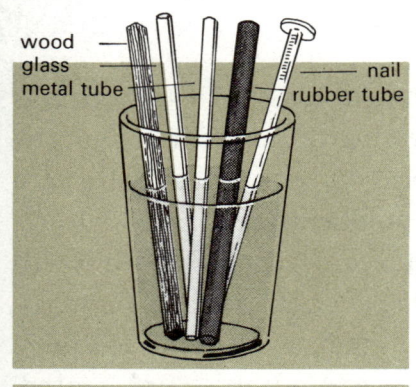

wood
glass
metal tube
nail
rubber tube

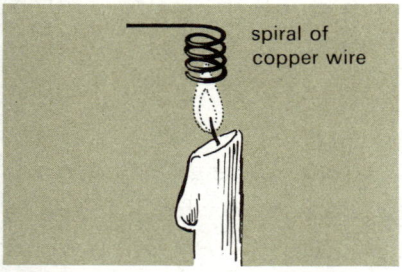

spiral of copper wire

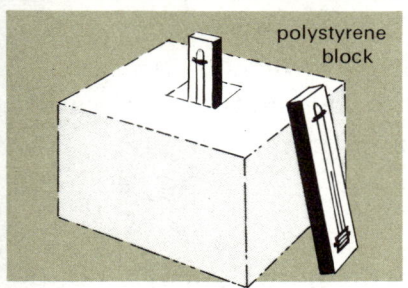

polystyrene block

Heat conductors

Some materials, especially metals, allow heat to pass through them very easily. They are bad heat insulators. They are good *conductors* of heat.

1 Put some very hot water (NOT boiling) in a can and in a plastic beaker. Try to pick up one in each hand.

Put tubes and rods of different kinds of metal, plastic, wood, rubber, etc., into a beaker of hot water as shown. By feeling the tops, see if you can pick out the good conductors of heat.

2 Wrap some bare copper wire round a pencil to form a spiral. Slide it off the pencil and lower it over a candle flame as shown. The copper conducts heat away so rapidly that the flame goes out!

3 Metals almost always feel cold when you touch them, because they conduct heat away from your fingers quickly. Insulators do not conduct heat away, and so they feel warm. A piece of expanded polystyrene actually feels as if it is *warmer* than other things in the room.

Make a hole in a block of polystyrene and put in a thermometer. What is the temperature of the block? Find the temperature of the room. What do you notice about the two temperatures?

Heat changes things

Evaporation

When substances are heated, a number of different things can happen.

If milk is heated slowly, water evaporates from it and it becomes a thicker liquid. This is how 'evaporated milk' is made. If the milk is now heated still more, even more water evaporates and in the end dry milk-powder is left. Instant coffee is made by dissolving in water all the soluble parts from ground-up coffee-beans and then heating the solution until all the water has evaporated, leaving behind the coffee powder. Often, fruit and vegetables are dried by gently heating them to evaporate the water from them. They keep for a long time in the dried condition without going bad.

Can you find out what fruits prunes and raisins were before they were dried?

Melting and boiling

Pouring molten iron

One of the simplest changes which heat makes to solid things is to melt them to form liquids. Cooking fat melts in a pan, the wax of a candle melts when the candle burns, ice melts in a warm room and solder melts on a soldering-iron.

Iron has to be heated to a very high temperature before it will melt. The picture shows molten iron being poured into moulds. When cooled, the original solid material is obtained again.

A similar kind of change takes place when liquids are heated. They boil and become vapours. If vapour is cooled, liquid is obtained again.

Try this experiment.

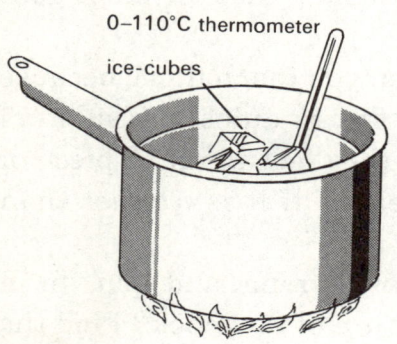

0–110°C thermometer

ice-cubes

Put some ice-cubes in a pan and heat them. Take great care not to scald yourself when the water gets very hot. Have a thermometer in the pan, one which reads from 0°C to 110°C.

Write down the temperature every half-minute.

Make a graph from your results like the one shown.

As long as there is some ice melting in the pan the temperature stays the same. What temperature is that?

Chemical changes

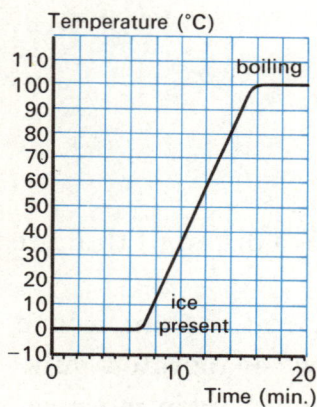

When you make a piece of toast you do more than drive water out of the bread. The outside of the toast goes brown, or even black if you heat it too much.

The heat has actually changed the bread to something different. This kind of change often takes place when we cook things. For example, when we bake a cake or cook meat, the heat changes the ingredients we start with to something quite different. This is called a *chemical change*.

What changes do you notice in an egg when you boil or fry it? When heat causes a *chemical change* you cannot get back the original substance by cooling.

Experiments with ice-cubes

1 How long does it take an ice-cube to melt in your class-room? Have a competition with some of your friends to see who can keep an ice-cube the longest. You could, for example, try packing it round with wool, cotton, sand, sawdust, aluminium foil and all kinds of material.

The record is supposed to be about 16 hours.

2 How can you make an ice-cube melt faster?

(a) Measure how long it takes an ice-cube to melt in a cup of water at 10°C, a cup of water at 20°C and a cup of water at 30°C.

(b) Does crushing the ice speed melting? Arrange an experiment yourself.

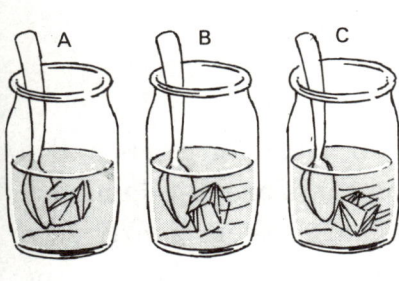

(c) Does stirring speed melting? Have three jars exactly similar as shown in the picture.

Leave A alone. Stir B slowly. Stir C quickly. Which cube melts first?

3 Which melts more quickly, an ice-cube on a saucer or an ice-cube standing on a dry sponge?

4 Let some water stand in the room for a long time so that it will have the same temperature as the air. Then put one ice-cube in the water and stand another on a plate in the air. Which melts more quickly?

Some more questions

Here are some more questions for you to answer by doing experiments.

1. Does your mouth 'stand' hot things better or worse than your finger? Make a very hot drink. Sip it when it is *just* cool enough to do so. Then put a *clean* finger into the drink. Can you hold it there?

2. Does a blackened pan heat up more quickly than a polished one? To experiment, everything must be exactly the same, other than the black and shiny surfaces. You can start with two similar shiny pans and blacken one over the flame of a candle. Put the same amount of water in each. Heat each pan on the same heater for the same length of time.
What do you notice about the temperatures?

3. It is said that hot water will freeze faster than cold water. This hardly seems likely. If you can find space in a freezer, put a dish of hot water and a dish of cold water side by side and find out.

5 Finding out about plants

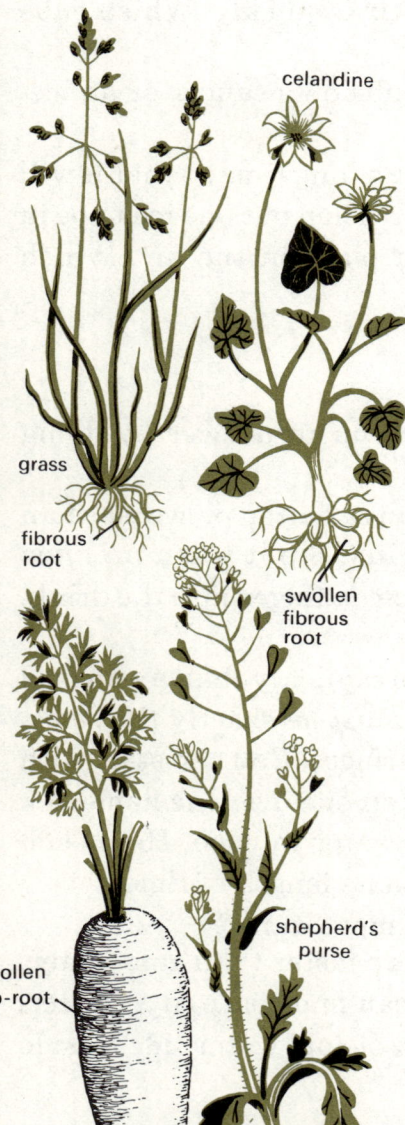

celandine

grass

fibrous root

swollen fibrous root

swollen tap-root

carrot

shepherd's purse

tap root

Look at roots

Carefully, with a garden fork, dig up a tuft of grass and a number of different weeds so that the roots are not damaged. Wash the soil from the roots by swishing them gently in a bucket of water. Look for two kinds of roots like those in the pictures.

What is the difference between the roots of grass and the roots of shepherd's purse? Roots like those of grass are called *fibrous roots*. Fibrous roots grow like a clump of threads. Roots like those of shepherd's purse are called *tap-roots*. Tap-roots have a main root with side branches.

Some plants have roots which store food. These roots become swollen. A carrot is a tap-root which has become swollen with stored-up food.

Find some celandine plants. If they are not flowering now, you may remember where they were in spring and be able to find the leaves. Dig up some roots. What do you see? Celandines store food in swollen fibrous roots.

What other food-storing roots do you know?

Ivy has strange roots which grow all along the stem. They cling to things as the plant climbs.

What is the longest root you can find? Dig up some dandelions and docks very carefully so that you do not break off the roots. Measure the lengths. What is the record?

The prize parsnips at garden shows are usually well over a metre long. Can you grow some as long as that?

Root hairs

Grow some mustard and cress in the way shown in the picture.

Sow mustard and cress seeds on damp cloth. Keep covered with a plastic bag.

foil tray

enlarged picture of root hairs

seed case

empty seed case

root hairs

daisy

plantain

leaf stalk

eye

underground stem

inside a crocus corm

Look at the roots of mustard and cress with a magnifying glass. Can you see the root hairs?

The plant takes in water through very tiny holes in the outside skin of the root hairs. Root hairs of other plants take moisture from the soil, too.

When the mustard and cress shoots are about 5 cm high you can cut them with scissors, wash them, and eat them. They taste good on bread and butter.

Look at stems

You will be able to find plenty of plants with round stems. Can you find some with square stems? Look on wet marshy ground for some with triangular stems. There are some plants with no stems at all. Two of these are daisies and plantains which you will find on most lawns. Can you find any others? These plants usually have a rosette of leaves growing near the ground.

Some stems are smooth and some are hairy. Collect twenty or thirty plant stems. Put the smooth ones in a glass of water and the hairy ones in another. How many are there in each set? How many hollow plant stems can you find?

Special stems

A potato plant has special stems for storing food. These stems stay underground and swell up in places to form the potatoes we eat. Dig up a potato plant carefully, so that the potatoes still hold on to the plant. See how the potatoes are formed on the plant.

Look at the underground stems of iris, bracken, lily of the valley, garden mint, coltsfoot, dog's mercury and Solomon's seal.

A crocus corm is simply the swollen base of the stem of the crocus plant. It is a store of food ready to give the plant a good start again next spring.

Look at leaves

Make a collection of different leaves. Show as many shapes as you can. Label each leaf with its name.

Collect one set of leaves which are all in one piece, like privet or beech, and another set of leaves which are made up of *leaflets*, like rose and ash.

Which is the smallest leaf you can find and which is the

Shapes of leaves

sycamore · ash · maple · wood anemone · poplar · oak · yew · maranta · pine · beech · willow · onion · tulip-tree · dandelion

biggest? Perhaps you could have a competition to find the biggest. If you get two leaves which are just about as big as each other, how are you going to decide on the winner? The picture shows you one way.

If you lived in South America, you might find water-lily leaves two metres across!

Edges of leaves

wood anemone · white dead nettle · dog-violet · oak · privet · holly

Giant water-lilies

Look at the edges of leaves. How many different kinds of edges can you find? Collect sets of smooth leaves, hairy leaves and leaves with prickles or stings.

Look for the veins in leaves. The veins are tubes which carry liquids around in the leaf. Some veins are straight, some divide into branches. Find leaves with each kind.

Veins in leaves

ivy · sycamore · lily of the valley · beech · tulip

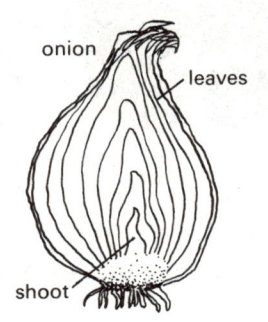

onion

leaves

shoot

Make a collection of leaves which have a strong scent when you pinch them.

Make a collection of *variegated* leaves. These are the leaves with markings and patterns on them in different colours. Grow some house-plants with pretty leaves. Could you arrange a house-plant show in your class-room?

Special leaves

Some plants have special leaves for storing up food over the winter. These leaves are packed closely together to make a *bulb*. An onion is a good example. Cut an onion down the middle to see the collection of fat leaves full of food.

Other special leaves help some plants to climb. They are called *tendrils*. Look for tendrils on pea plants and clematis.

pea plant

tendrils

Many animals eat leaves

Cows, sheep and horses and many other animals feed on grass leaves. Farmers cut grass and dry it to make hay so that these animals have food in the winter.

We grow cabbages, lettuces and similar plants for leaves to eat ourselves. Slugs, snails and some kinds of caterpillars like these, too.

Other kinds of caterpillars eat the leaves of nettles, goose-berries, oak-trees and very many other plants. Each kind of caterpillar eats one particular kind of leaf.

Some leaves eat animals

sundew

butterwort

Sundew and butterwort grow in Britain on marshy moorlands. Flies stick to their leaves. When a fly is caught the leaf closes over it and gives out a fluid which dissolves the fly. Pitcher plants grow in bogs in Malaysia. The 'pitcher' forms at the end of a leaf and collects water. Insects fall in and are digested by the plant. Why are insects attracted to the pitcher, do you think?

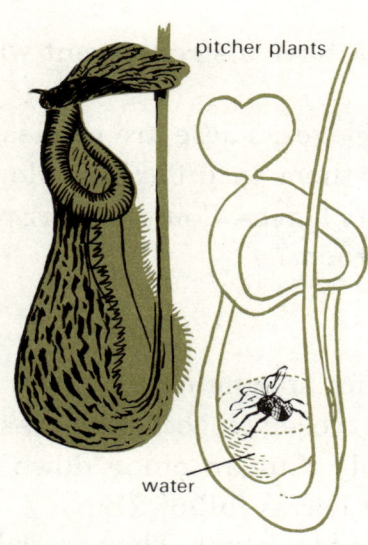

pitcher plants

water

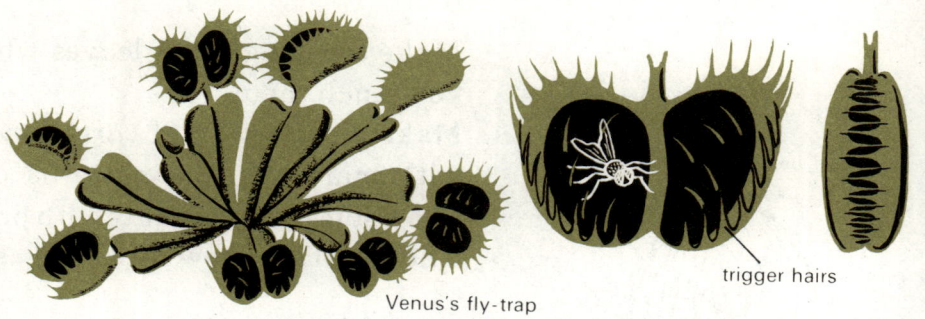

Venus's fly-trap

trigger hairs

The leaves of the Venus's fly-trap can move quite quickly. Whenever an insect touches the trigger hairs, the two parts of the leaf close on it like a trap.

Another plant which moves quite quickly is the sensitive plant. The leaves droop if you touch them and then slowly open up again. You can often buy plants of Venus's fly-trap and the sensitive plant in pots, and they are quite easy to grow in the house.

sensitive plant (mimosa)

Look at flowers

Collecting

During spring and summer there are a great number of wild flowers to find. A good way to learn their names is to collect six different ones each week. Pick only one or two of each flower and put them with some of their leaves in separate small jars. Put the name in front of each jar, on a piece of paper or card. You could show garden flowers, too.

Making flower booklets

Make some booklets about the wild flowers which grow in certain places at certain times of the year. One might be called 'Flowers in the school field in May', another 'Roadside flowers in June and July', and so on. In your booklet have coloured drawings and some pressed flowers with their names. Other people who find the flowers will be able to look up the names of them in your book.

sunflower

speedwell

The size and colour of flowers

What is the largest flower you can find? What is the smallest? Make collections of flowers of one colour. One group could collect red flowers, another group yellow flowers, and so on.

Parts of a flower

Look closely at some of the flowers with a magnifying glass. Look at the different parts. Look for those shown in the picture. (Some flowers do not have all of them.)

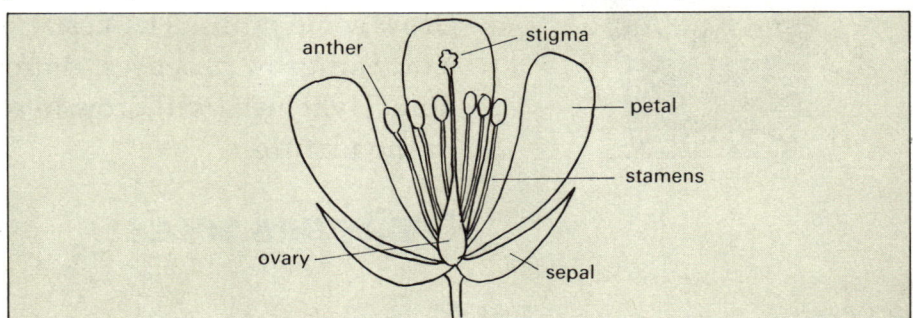

When anthers are ripe, they burst to let out tiny grains of pollen. Can you see some through your magnifying glass? Pollen gets carried by insects or the wind. When some pollen lands on the stigma of a flower of the same kind, then this flower is *pollinated* and the seeds in the seed box will begin to grow.

Flower families

Two members of the pea family

Look for family likenesses between flowers. For example, look for the 'cross family' which all have four petals in the form of a cross, the pea family which all have flowers with the same arrangement of petals as the sweet pea, and the daisy and dandelion family whose flower heads are made up of a lot of separate small flowers growing together.

Plants without flowers

There are several kinds of plants which do not have any flowers at all. Ferns, mosses, fungi and seaweeds are examples. We shall be looking at these plants in Book 4.

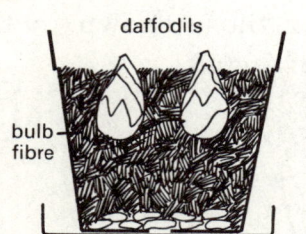

How deep to plant

crocuses — bulb fibre

daffodils — bulb fibre

Growing bulbs

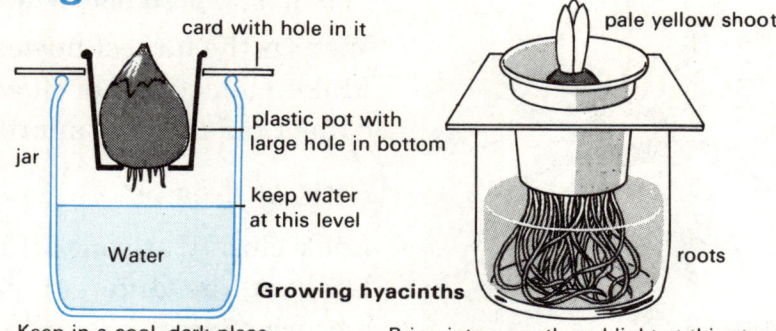

card with hole in it

jar

plastic pot with large hole in bottom

keep water at this level

Water

pale yellow shoot

roots

Growing hyacinths

Keep in a cool, dark place

Bring into warmth and light at this stage

Grow some bulbs. The best time to plant them is in autumn. You can grow crocuses, daffodils and tulips in pots of bulb fibre. Hyacinths will grow in water. You can do this as shown in the picture.

Fruits and seeds

water-melon

apple

The pips in an apple core are apple-tree seeds. See if you can grow some of them in a pot of soil. Look for the seeds in all the kinds of fruit you eat. A fruit is a ripe seed box. Not all of them are good to eat. Some are dry and hard, and some berries are poisonous. Some fruits have a stone inside. Crack the stone open and you will find the seed inside.

Look at a gooseberry. Gooseberries have 'tops' and 'tails'. The top is the remains of the petals of the flower, and the tail is the flower stalk. All fruits have *two* marks on them like this. Seeds have only one mark which shows the place where they were fixed on to the fruit.

tomato

gooseberry

raspberry

seeds on outside

strawberry

hazel nuts

elderberries

rose-hips

hawthorn haws

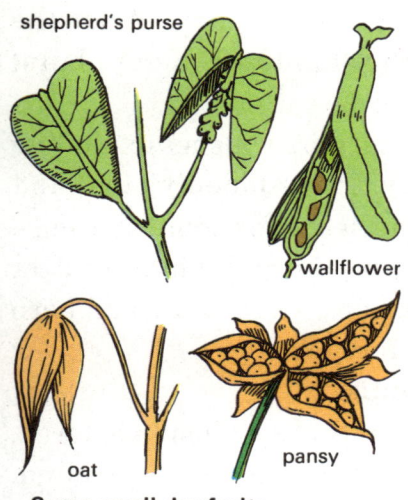

shepherd's purse

wallflower

oat

pansy

Some small dry fruits

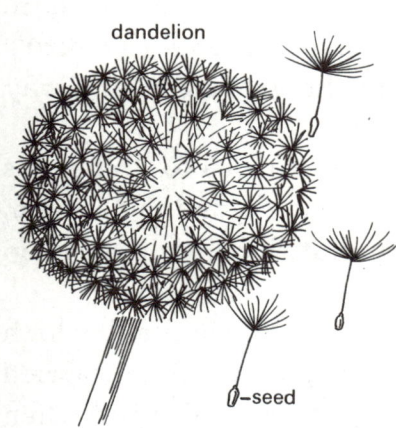

dandelion

seed

poppy seed box

seed pod

seed

policeman's
helmet
(Indian balsam)

seed pod
bursting

How many different fruits can your class collect? Find the seeds in all of them. Plant as many kinds of the seeds as you can find room for. See how they grow. Who can grow the tallest sunflower, the heaviest marrow and the longest runner bean?

How seeds travel

Seeds must somehow get away from the parent plant. They must have room to grow themselves without being too close together. How do they travel? Some seeds are carried by the wind or by animals, some are scattered by the plant itself.

1 **Wind travellers**

In summer and autumn collect fruits and seeds which travel by wind. Often it is the whole fruit which travels and it has special attachments to help it float through the air. Some, like dandelion, willow-herb, clematis and groundsel, have parachutes made of fluffy hairs. Others, like sycamore, ash and lime have wings.

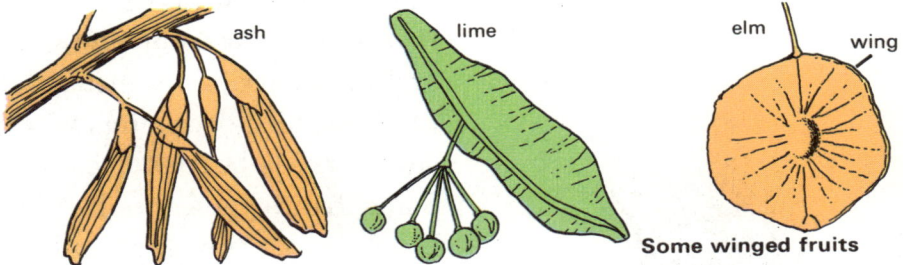

ash

lime

elm

wing

Some winged fruits

Try this experiment. Stand on a chair and hold up some winged seeds as high as you can. Let them fall through the air. How do they go? How far away from you do they travel?
Measure the longest distance for each kind of winged seed.
Do the same experiment with some 'parachute' seeds. You need not climb on a chair, just blow them off the palm of your hand. The wind makes poppy seeds travel in another way. The seeds are very small and are contained in a box on the top of a stem. You will find them in September. The box has holes in it all around the top, and when the wind shakes the stem the seeds fly out for quite a distance.
Look for snapdragon seed boxes in late summertime. Snapdragons scatter seeds like the poppy does, and so does the springtime wild flower called campion.

2 **Seeds which spread by explosion**

If you look near the edges of streams in late summertime, you may find the plant in the picture. It is between one and two metres tall and has hollow stems. If you hold a flower with the

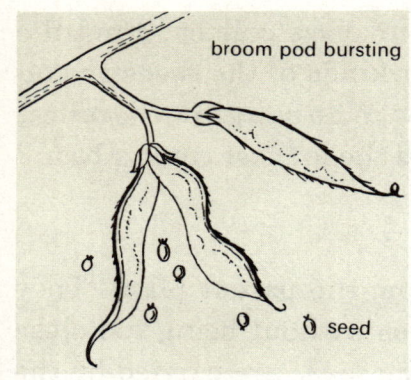

broom pod bursting

seed

stalk upwards, you will see how it gets its nickname. When the seed pods are ripe they only need the slightest touch to burst open with a 'pop' and scatter the seeds.

Have you heard the pods on broom and gorse bushes exploding on a hot day in summer? As they dry, they suddenly burst and twist at the same time, throwing the seeds for some distance. How many more kinds of plants can you find which scatter their seeds by small explosions? Pansies, sweet peas and herb robert are interesting ones to look for.

3 Seeds which are carried by animals

Many fruits are carried off by birds. If the seed inside is large, the birds eat the juicy part and drop the seed on the ground. They do this with cherries and hawthorn haws. If the fruits have small seeds like blackberries, raspberries and elderberries, birds swallow them whole. But the seeds are so hard that they are not digested. They pass through the bird's digestive system and are dropped on the ground a long way from the parent plant.

If you live near moorland you will be able to look for bilberries and crowberries in autumn. Look for the seeds inside them. Look for bird droppings which show that the birds have been eating these berries. How can you tell?

Some fruits have hooks. These cling to any animal which brushes past the plant and the fruit gets pulled off and carried away. The animal may rub it off or it may simply drop off when the hooks dry up and shrivel. Look for burdock on waste ground and waysides in summer, and goose-grass (cleavers) in hedgerows in spring.

burdock seed head

Mice and squirrels disperse seeds another way. They make little stores of fruit stones and nuts to eat in the winter. They do not eat them all, and the 'left-overs' grow into new plants.

Some experiments with plants
Collecting seeds

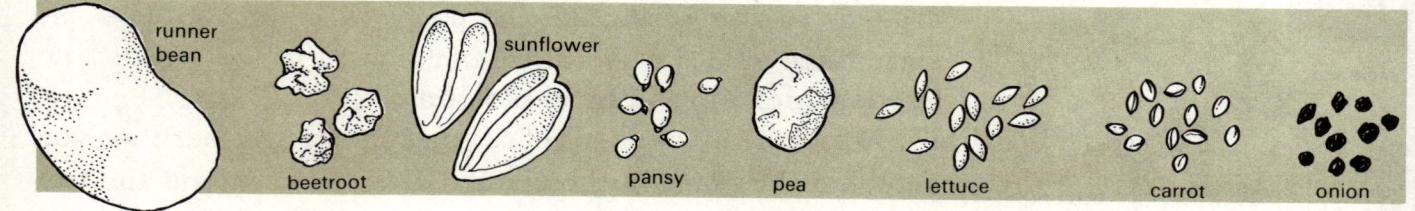

runner bean

beetroot

sunflower

pansy

pea

lettuce

carrot

onion

Make a collection of different seeds. Collect from wild plants and from garden plants during the summer and autumn. Make sure the seeds are ripe. You can also collect seeds from packets sold in garden shops. Shops usually begin displays of seed packets in January. Label the different kinds with their names. Look at the different shapes. Which is the biggest and which the smallest?

Look inside a broad bean seed. To do this you will need to soak several seeds in water overnight. When the seeds are well soaked, squeeze one of them. What do you notice? With a needle, take off the tough outside skin. What do you find inside? Make a drawing like the one below.

What do you think the two large 'hinged' parts are for? Look inside some other large seeds in the same way.

A broad bean seed

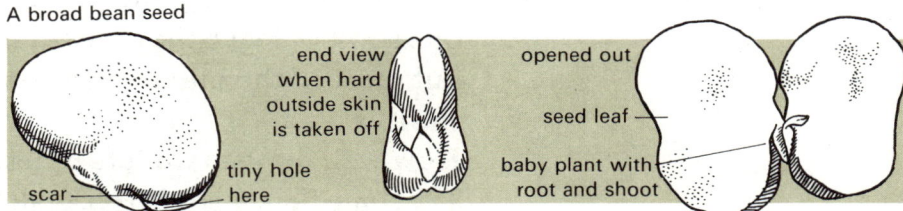

Growing seeds

When seeds begin to sprout, we say they are *germinating*. They need three things to germinate: warmth, moisture and air. Why don't seeds germinate in the packets in shops?

Choose a few kinds of seeds and try to grow them as shown in the picture.

Keep them in a warm room.

See that each pot is labelled, and count the number of seeds planted in each one. Make a chart like this and fill it in for all the kinds of seeds.

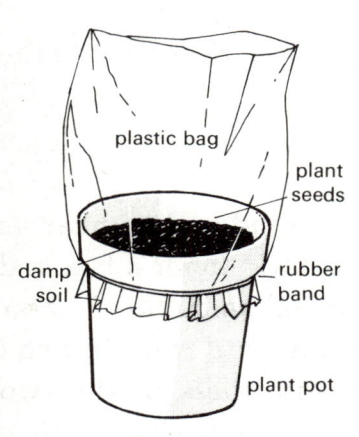

Name of seed	No. of days before 1st. sprout seen	No. of seeds planted	No. which germinate

Put one lot in a refrigerator so that you can compare them with those grown in the warmth.

Does it matter how deep you plant seeds?

Find a deep box or plant pot and plant some seeds in it at different depths. Put 2 cm of soil in the bottom and plant a seed on this. Then fill in another 2 cm of soil and plant the second seed and so on. Keep the soil moist. What happens? Try the experiment again with a different kind of seed.

Does it matter which way up you plant seeds?

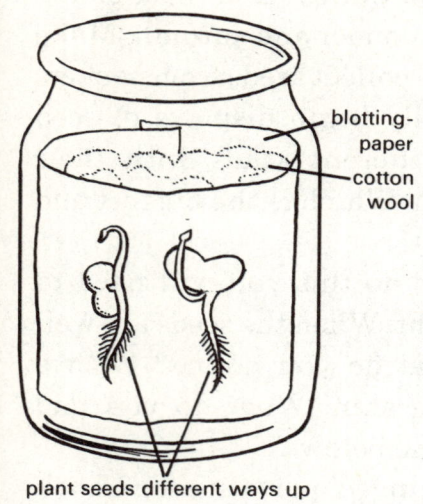

plant seeds different ways up

The picture shows how you can put seeds in different positions to germinate. You can use peas, French beans or marrow seeds. Make sure that the mark on each seed points in a different direction.

Make some drawings to show how the roots and shoots come out of the seeds and how they grow.

Plant seed potatoes in the garden or in pots. Plant them different ways up. Is there any difference in the way they grow? Will it matter which way up you plant a daffodil bulb?

Plants need space

Sow two rows of lettuce seeds. Follow the instructions on the packet. When the plants have grown about 5 cm high, thin out one row but leave the other alone. Thinning out means pulling out most of the plants, just leaving one every 10–15 cm along the row.

How well do the two lots of lettuce grow?

Try some flower seeds too, godetia or clarkia are good kinds to try. Sow half of a packet of seeds very close together.

Patiently plant the others about 10 cm apart. How well do the two lots grow and flower?

Plants and light

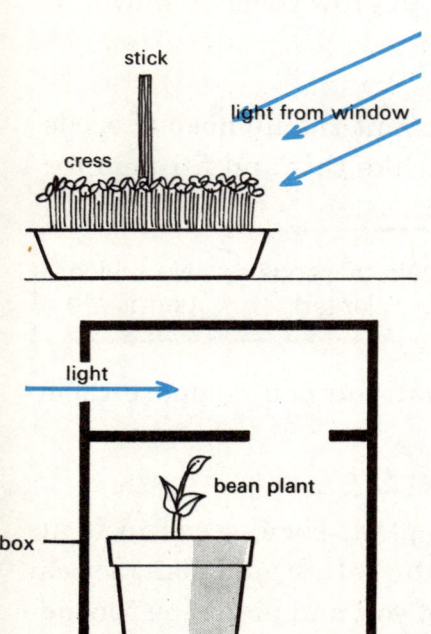

1 Get two healthy pot plants of the same kind and as near as possible the same size. Keep one in the dark under a box and the other in the light. Give them both equal amounts of water and keep them at the same temperature. Write down what you notice about the colour, height and sturdiness of the two plants after they have grown like this for a few weeks.

2 Grow some cress seeds as shown in the picture. Give them plenty of light and keep turning the dish so that they grow straight up. Now put the dish in a sunny window so that the light comes all from one side. Write down the time. Look at the cress from time to time and make a note of when you first notice a bend of the stems towards the light. What is the stick for?

Try the experiment with a table-lamp instead of sunlight. Does it work? Do the plants bend quicker or slower than before?

3 Grow a bean plant or a pea plant in a pot, and then put it in a box like the one in the picture. Keep it watered. What happens?

Measure how quickly some plants grow

When you grow a bulb in a pot, measure the height of the shoot each day. The easiest way is to cut a piece of string the same length as the shoot.

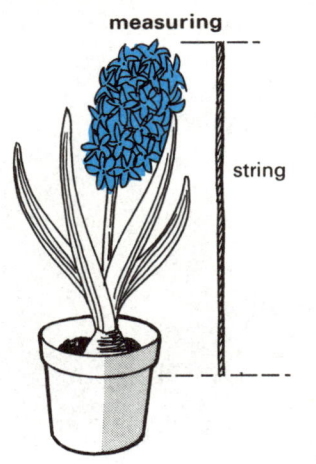

measuring

string

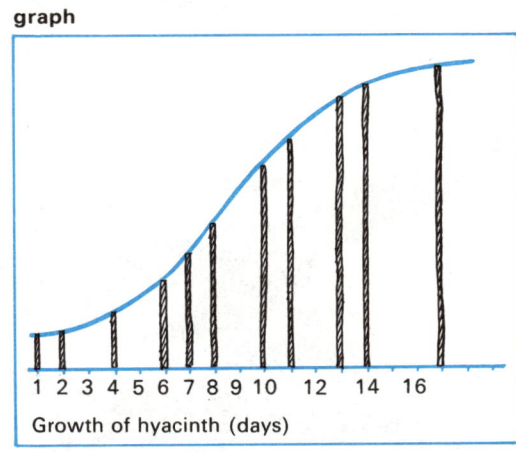

graph

Growth of hyacinth (days)

If you stick the pieces of string on to paper as shown in the picture you will make a growth graph. What do you notice about its shape? Do this for some other plants growing in the garden in spring and summer. How fast do the new shoots on rose-bushes grow? How fast does the shoot of a vegetable marrow grow? What is the fastest growing shoot you can find? Find some plants which grow only slowly.

Shoots can push up with great force

Have you ever seen plants growing up through asphalt paths? Do they grow through a crack that was there already or do they push through? Grow some beans or a small dandelion in a pot of soil. Put a box on top, with a small stone on it, as shown in the picture. Do the plants lift the stone? If they do, try a heavier one.

How heavy a stone will they lift up?

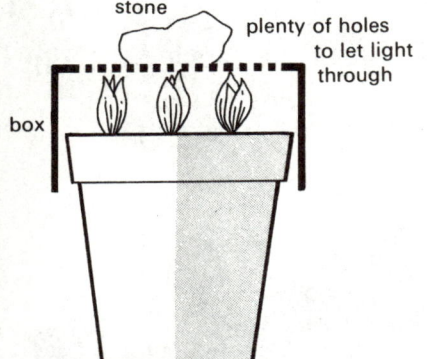

stone

plenty of holes to let light through

box

6 Birds

Birds in towns

house-sparrow (cock)

chaffinch

house-sparrow (hen)

robin

great tit

blue tit

starling

male blackbird

thrush

magpie

rook

pigeon

You can learn a lot about birds by watching them in your own garden, in the school grounds or in a park. Even in towns there are a number of different kinds of birds. Can you spot all those in the following list?

House-sparrow, chaffinch, starling, hedge-sparrow, robin, blackbird, thrush, wren, blue tit, great tit, pigeon.

In winter the black-headed gull may come into towns. In summer you could possibly have house-martins, swallows and swifts. Town parks sometimes have lakes with ducks and perhaps moorhens on them.

Fix up a bird-table where you can watch it from a window. Do not fix it near to shrubs where a cat can hide.

Feed the birds on your table in winter-time. On the side, you can hang meat bones, strips of bacon rind, some peanuts in a net bag or perhaps half a coconut. On the top, put bread-crumbs, soaked dog biscuit and almost any kind of kitchen scraps. Which food does each kind of bird seem to like best? Which birds are shy and which are bold? How do the different birds perch on the table? Which is the last one to come in the evening?

You should not feed the birds between April and October. There is plenty of food for them then to get without your help, and natural food is much better for the young birds in the nests than our scraps. It is important to provide birds with a shallow dish of water all the year round. In very dry weather, and in frosty weather, they find it especially hard to find water in towns. Watch exactly how a bird drinks.

Can you find out which birds open milk-bottle tops? How do they do it? Do the old ones teach the young ones what to do? Watch starlings on the grass. How do they use their beaks? They are obviously getting food. What can the food be? Cut out a piece of turf and put it on some old newspaper. Pull apart the grass roots. Can you find the grubs that the starlings are eating?

Birds in the country

Most of us can look for birds in the country as well as in the town. How many birds do you see in country places? There are many more than in towns. Keep a record of those you observe, noting the date, the place and any interesting details. When you are by the sea, look for the birds shown in the picture.

black-backed gull

tern

herring-gull

oyster-catchers

gannet

puffin

black-headed gull

kittiwake

Owls, rooks, magpies, woodpeckers, pheasants, wood-pigeons and tree-creepers shelter in trees and bushes. Find these birds in the picture.

wood-pigeon

od–
ker

little owl

tree-creeper

rooks

magpie

pheasant

Birds which live in moorland areas often make their nests on the ground. The curlew lives for most of the year on the mud of river estuaries, but flies inland to the moors to make its nest.

Some moorland birds are shown here

grouse

wheatear

meadow pipit

skylark

curlew

swan

mallards

moorhen

dipper

kingfisher

Some birds live only near fresh water in lakes, reservoirs, ponds and streams.

What to look for

When you spot a bird, there are several questions to ask yourself.

1 **How big is it?**

Get to know the size of three common birds; a rook, a blackbird and a sparrow. They are drawn to the same scale in the picture.

rook

blackbird

sparrow

Then when you see another bird, ask yourself how it compares with these.

2 What shape is it?

Is it round and fat, or is it slim? What shape is its tail? Is the tail short or long and how does the bird hold it? Make sketches in a notebook to show the answers to these questions.

swallow

peacock

pheasant

woodpecker

wren

thrush

magpie

Look at the different shapes and sizes of the birds in the picture and at their different tails.

Does it keep its tail still, or move it? Look out for a wagtail, to see if it deserves its name.

3 How does it stand?

Does it hold its back fairly upright, or level, or somewhere between the two positions? Does it often stand on the ground or does it perch on something, such as a twig or a fence? Try to see what its feet are like. The next picture shows some different birds' feet. In what different ways are these feet used?

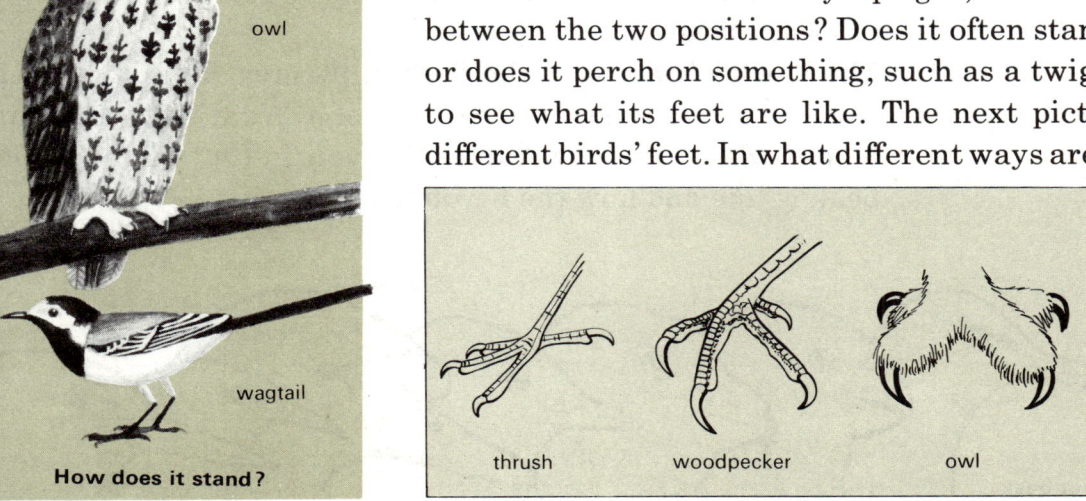

owl

wagtail

How does it stand?

thrush

woodpecker

owl

mallard

blue tit

thrush

chaffinch

goldfinch

bullfinch

4 What colour is it?

Some birds are black, some are dull coloured and well camouflaged in the places they live. Others have bright colours and special markings.

Notice the colours and the markings on birds and where they are. Are they on the breast, back, head, wings or tail? Look for white patches. Are there any stripes behind the eye? Keep all the answers in your head and then, when the bird has flown away, get out your notebook and make a sketch while you can remember. When you get home or back to school make a better drawing and colour it.

5 What sort of beak does it have?

Is it narrow or broad? Is it long or short? Is it straight or curved?

Birds with straight short beaks, like the finches, use them for cracking seeds. Fine pointed beaks are the best for the insect-eating birds. The woodpecker's beak is pointed and strong for hammering holes in wood and for prising grubs out of cracks. The curlew's strange beak is used to obtain food out of the mud of river estuaries.

The hawk's beak tears the flesh of its prey. The duck's beak scoops up water and filters food out of it. What do thrushes and blackbirds use their beaks for? Can you find out what a heron's beak is like and how the heron uses it?

Some different beaks

6 How does it move on the ground?

Some birds hardly ever land on the ground. If a swallow does come to the ground, it has difficulty in taking off again. Watch birds which move along the ground. Do they hop with both feet together, do they walk or do they run?

Have they more than one way of moving along?

Why does a blackbird on a lawn keep turning its head on one side?

Why does a wagtail keep darting this way and that?

7 What sort of song does it sing?

Listen to birds singing. Can you imitate some of them? At what times of the day do birds sing most? In what seasons of the year do they sing? Which birds have more than one song?

Listen for the warning cry of blackbirds and thrushes when a cat is about, or when you frighten them. Do other birds have warning cries?

8 How does it fly?

Does it fly in a steady level line or does it rise and fall as it flies? How fast do the wings beat? Does it swoop and dive through the air? Does it glide with wings almost still?

Does it hover, staying in the same position with its wings beating? Can you think of a bird which flies vertically upwards from the ground?

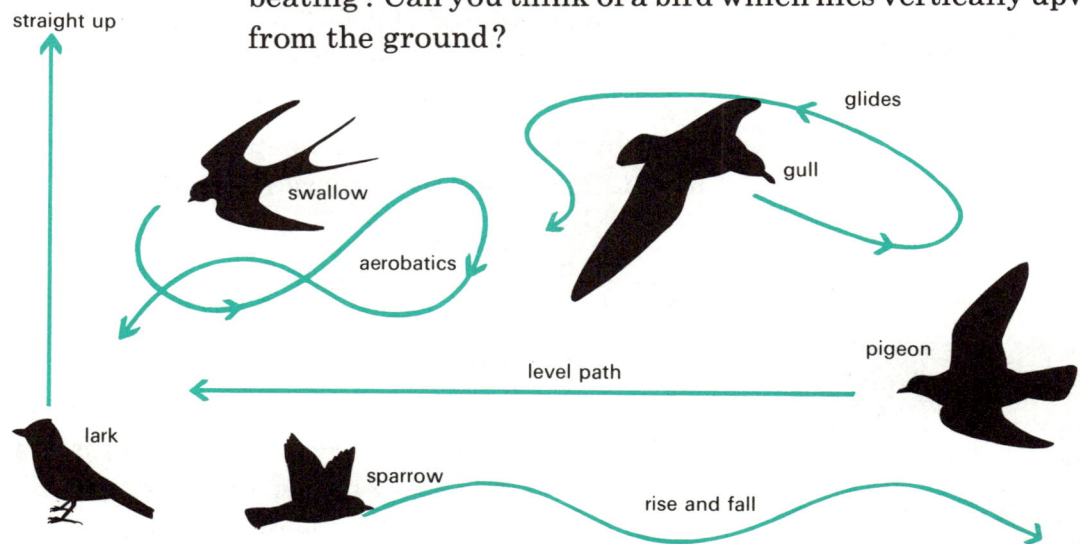

Look at wing shapes as birds fly, and make some silhouettes like those in the picture.

Feathers

A down feather.
Young birds and some
adults (e.g. ducks)
have these.

A wing feather

Make a collection of different kinds of feathers. How many colours and patterns can you find? Some feathers are called *coverts*. These are the feathers which simply cover up parts of the bird. They are often fluffed out in winter to keep the bird warm by enclosing a layer of air. The most interesting feathers are the *wing* feathers. They have a stiff central part called a *quill*. The quill is hollow, so that although it is strong it is also very light. Make yourself a quill-pen as shown in the picture. Can you write with it? When were pens like this used? Look at a wing feather through a microscope. Try to pick out the different parts shown in the picture.

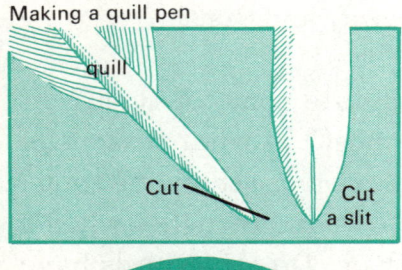

Making a quill pen

quill

Cut

Cut
a slit

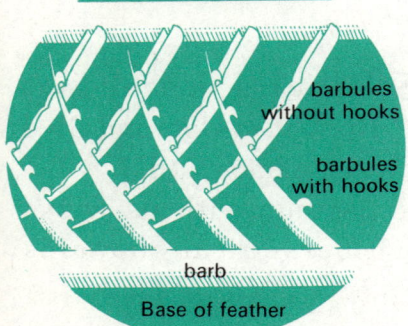

Tip of feather

barbules
without hooks

barbules
with hooks

barb

Base of feather

Magnified view of wing feather
looked at from underneath

As the bird pushes its wings downwards when it is flying, the little hooks hold the barbules together to press on the air. When the wing moves upwards, the hooks lift off and allow the barbules to come apart so that the air can flow through them.

Feathers grow from a bird's skin like hairs grow from our skin. Sometimes some of a bird's feathers fall out and new ones grow. This is called *moulting*. The feathers you pick up for your collection are likely to be those that have moulted. If you have a pet budgerigar or if you keep hens, you will have seen them moult. They look very sorry for themselves for a few days.

Nests

House-martins build mud nests under
the eaves.

Most birds build nests. Do you know one which does not? Each bird makes its own kind of nest. Some use moss, feathers and dried grass, others use mud or even just a few stones on a shingle beach. Each kind of bird has a favourite place for making its nest. It might be on the ground, in a hedge, in the fork of a tree, on a ledge in a barn, under the eaves of a house or in a hole in a bank. A blackbird, for example, builds a grass nest in bushes. A robin makes a little nest of moss, feathers and hair, often in old plant pots or garden rubbish. You can watch house-martins building very easily, because their nests are not hidden. They make them completely of mud. They have to carry every bit of mud in their beaks from the nearest pond,

A moorhen's nest is hidden amongst reeds.

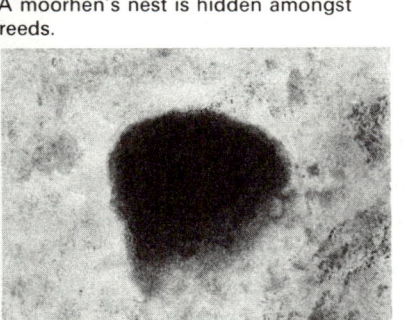

Sand-martins make a deep hole in a sandy bank for a nest.

A thrush's nest is lined with mud.

A tern makes a rough nest on a rocky ledge.

stream or ditch. Then they stick the mud pellets to the wall and eaves of a house with saliva.

Birds do not live in nests as though they were homes. Nests are built just to hold the eggs and to rear the young birds in until they get their feathers and can begin to fly.

Nesting materials

In April and May, put out different kinds of nesting materials in a place where you can watch from a window. You might have straw, dried grass, wool, string, feathers, hair, bits of cloth and moss.

Watch which materials the different birds choose to take.

Watching nests

During the spring and summer see how many nests you can spot. *Watch but do not touch*. Stand some distance away to watch what the birds are doing. NEVER FRIGHTEN THE BIRDS AND NEVER TAKE EGGS. Only go near to look at the nest when you think that *both* parent birds are away and then do not go very often. If you do, you will break or move leaves and twigs and make it obvious where the nest is. The birds are almost certain to desert the nest if you look too often. You may be able to see how many eggs there are, how big they are and their colours. Make a coloured drawing. How long does it take for the eggs to hatch? You may be able to get a quick peep at the baby birds while the parents are away. Have they any feathers? What do you notice about their mouths? Watch the parent birds as they come back with food for the young ones. See if you can tell what they are feeding the young on. How many journeys do the parents make in ten minutes? How many will that be from daylight to dusk? No wonder young birds grow quickly!

Always protect nesting birds. Perhaps you could fix a safe nesting-box in your garden.

A nesting-box for tits

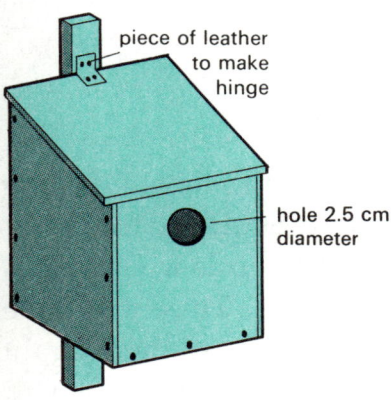

piece of leather to make hinge

hole 2.5 cm diameter

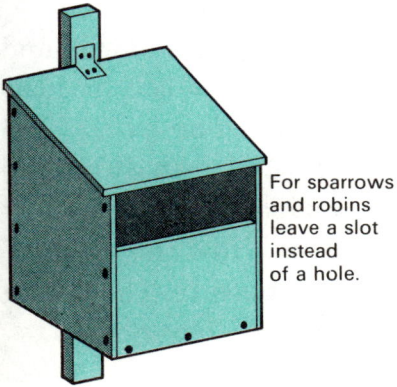

For sparrows and robins leave a slot instead of a hole.

Fledglings

Watch out for young birds which have grown their first feathers and have come out of the nest. They are called fledglings. This is a very dangerous time for them because they can hardly fly at all. Try to keep cats away from them. The parent birds go on feeding them for quite a time. The young birds often make loud noises and flutter their wings to beg for food.

Old nests

When birds have quite finished with a nest you can collect it to look at. Take it to pieces and arrange the materials in little piles of the different kinds to show what the birds used. Can you spot any bits that you put out in the spring?

Migration

Swallows gather together before they migrate

House-martins and swallows are summer visitors to this country. Every summer they come back to their old nesting places after being away for five months in South Africa. What an amazing power of flight they have. The birds we watch in summer fly away in September or October. They will have travelled nearly 20 000 kilometres before we see them again next spring. That is about twice as far as most families will have driven in their cars during the same period!

Two other summer visitors to this country are the swift and the cuckoo. Make a note of the date you first see or hear them, and also when you think they have gone away. The older parent cuckoos fly away to central Africa in July. The young ones are not fully grown then, but later, in August or September, they also fly to the same place. How they 'know' where to go is quite a puzzle. Swifts do the opposite. The young ones fly off to central Africa first, before their parents.

7

Grasses and small animals

Grasses

annual meadow grass   oat grass   fescue

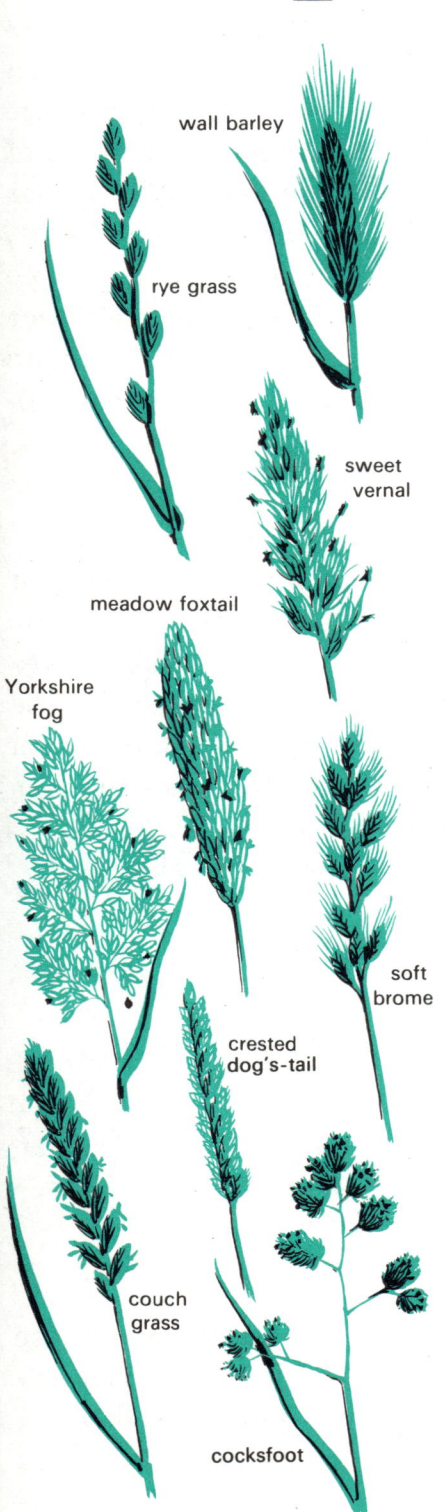

wall barley

rye grass

sweet vernal

meadow foxtail

Yorkshire fog

soft brome

crested dog's-tail

couch grass

cocksfoot

Many people think of 'grass' as a plant which makes lawns and playing-fields and which needs a lot of work to keep it short and free from weeds. But there are many different kinds of grasses. If they are not cut down they will grow quite tall and will have flowers and seeds.

May, June and July are good months to study grasses. Most of them are flowering then. See how many different kinds of flowering grasses you can collect.

You may not be able to name them all, but the pictures will help you with a few.

Grasses in different places

Look at waste-land, woodland, meadow land, moorland, road-side verges and the edges of playing-fields. Make separate collections of grasses from each place. How are they different? Are there some kinds of grass which are able to grow almost anywhere?

There are some other kinds of plants which often grow along with grasses. Each month from May to September, make a list of the wild flowers you find growing in a meadow or other grassy place.

In a field where animals have been grazing, look at the plants they have avoided and make a list. Why do you think these plants have not been eaten? Is it the taste or smell which the animals do not like? Or do the animals find the plants unpleasant to touch? Watch horses, cows and sheep eating grass. How does each one eat? Which of them swallow the grass and then chew the cud later on? What does the grass in the field look like after each kind of animal has been grazing for a time?

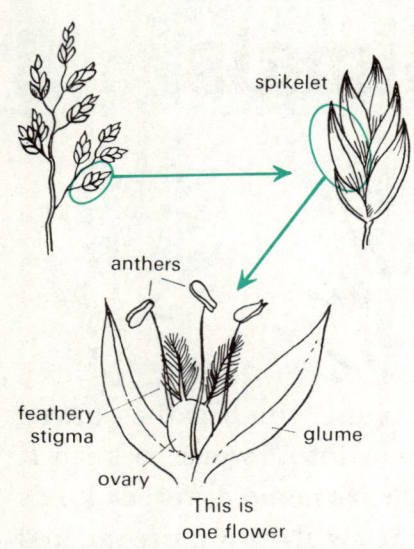

spikelet

anthers

feathery stigma

ovary

glume

This is one flower

A closer look at grass flowers

Now use a ×15 or ×20 microscope, or a good magnifying glass, to have a closer look at some grass flowers. Choose those which have opened widely. The drawing will help you to find out what the different parts are called. The flowers are small and not very showy. They do not have to attract insects to carry pollen from one to another. The wind does this. Take a grass plant with the flowers wide open so that the anthers are showing. Give the stalk a quick shake to see what a lot of pollen comes out. Even a slight puff of wind shakes the slender stems of the grasses, scatters pollen and carries it far and wide. Look at the anthers closely with your magnifier. Can you see any pollen grains? Poke the anthers with a needle and notice how loosely they are joined to their stalks. These stalks are very slender, too. You can see how well the anthers will shake in a wind.

Grasses provide food

Grasses are very important plants because they provide most of our food. We eat the seeds of several kinds of grasses ourselves, and also feed grasses to the livestock which gives us meat, milk and eggs. The seeds, stems and leaves can all be dried, and then they store very well to feed us and our farm animals during the winter. We often grind the seeds to make flour, or roll them flat to make breakfast flakes.

Learn to recognise the food grasses shown in the picture.

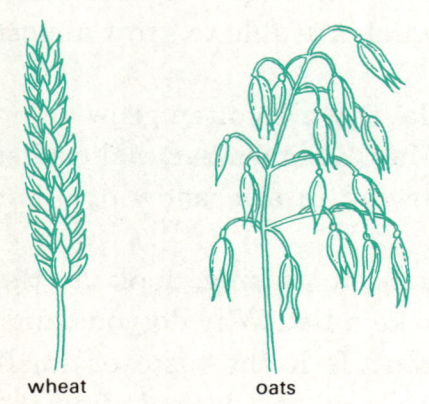

wheat   oats

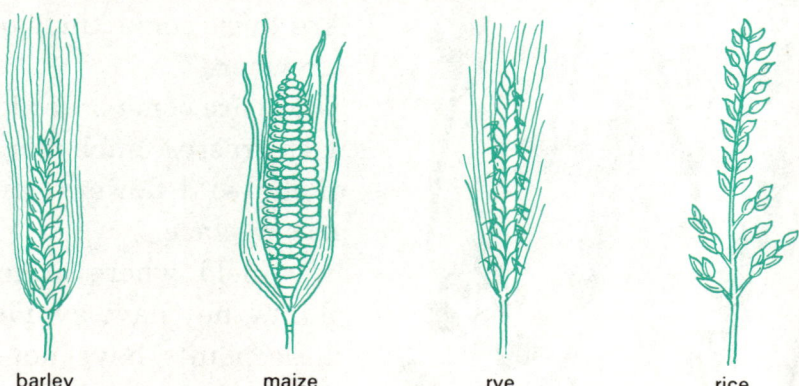

barley   maize   rye   rice

All the drawings are about ½ size except the maize which is about ¼ size.

Look at the fields in your district to see whether any of these crops are growing. If they are not grown in your district, try to find out why not. Which of them have you eaten today? Grow some of them in your garden. How many seeds do you get from one?

Some small animals

Grasshoppers

When you are looking at grasses on warm summer days, you may be able to catch some grasshoppers. You will have to be quick. A grasshopper has wings but cannot fly. The wings help the large back legs when the insect makes huge jumps. Try to find out how many times its own length a grasshopper can jump.

How many times your own height can *you* jump forwards? You can keep grasshoppers for a short time in a large glass or plastic container with plenty of grass. They eat grass and many other green plants.

There are two kinds, 'shorthorns' and 'longhorns'. You will see from the picture why they have these names.

Have you heard the chirping noise which grasshoppers make? Only the male makes this noise. The shorthorn kind makes the sound by rubbing the inside surfaces of his largest legs on his hard front wings.

The longhorn kind does it by rubbing his wings together.

Shorthorn grasshoppers have eardrums on the sides of their bodies. Longhorns have them on their front legs!

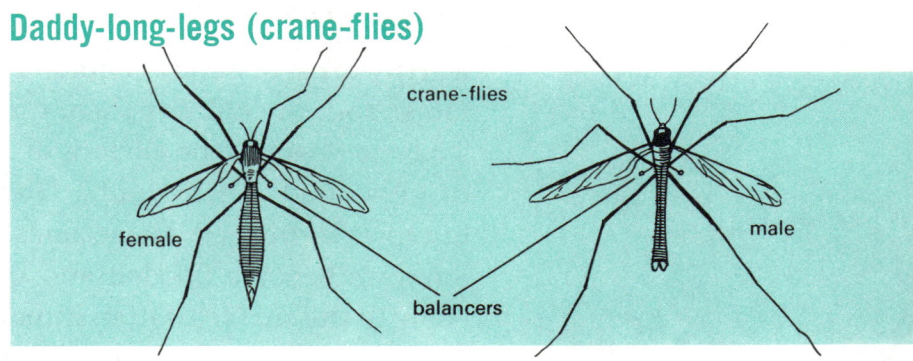

shorthorn grasshopper

longhorn grasshopper

Daddy-long-legs (crane-flies)

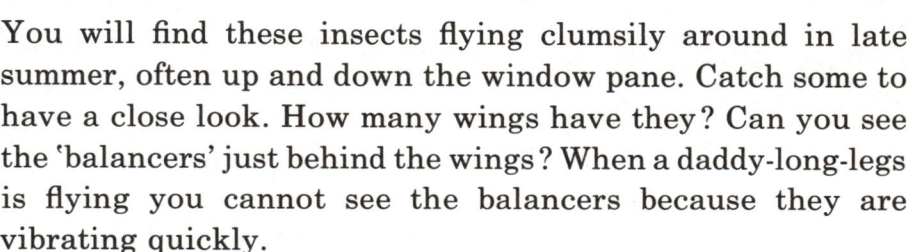

crane-flies

female

male

balancers

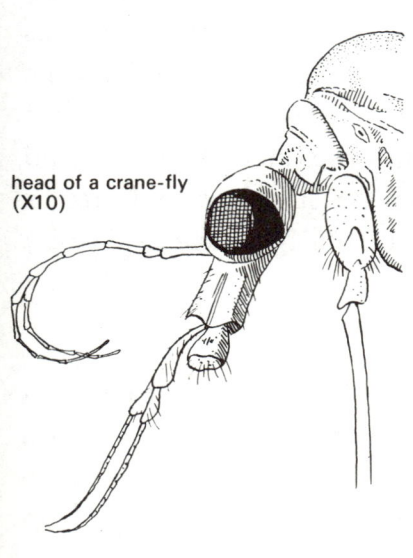

head of a crane-fly (X10)

You will find these insects flying clumsily around in late summer, often up and down the window pane. Catch some to have a close look. How many wings have they? Can you see the 'balancers' just behind the wings? When a daddy-long-legs is flying you cannot see the balancers because they are vibrating quickly.

Find some males and some females. The picture will show you how to tell the difference. Look at the rear end.

They have no jaws so they cannot bite. They like to sip at liquids. Put one near a juicy apple core and watch.

leather-jacket

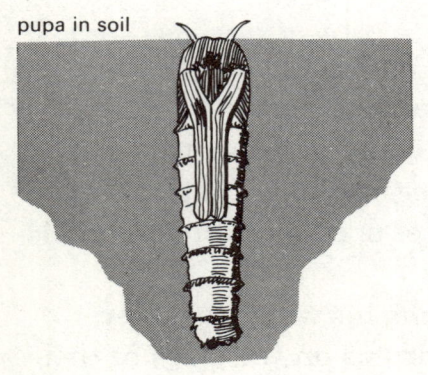
pupa in soil

earthworm

Eggs and larvae

Each female daddy-long-legs lays 200 to 300 eggs. Every time she settles on the grass she pushes a few in among the roots with the pointed end of her body. The eggs hatch into larvae which grow to 2–3 cm long. These are called leather-jackets because they have a tough greenish-grey skin. Leather-jackets eat grass roots, and sometimes there are so many of them that they kill off the grass over a large area.

Finding leather-jackets

Cut a piece of turf and pull the roots apart to find some leather-jackets. With a magnifying glass, you will probably be able to see their small head and jaws and the two breathing-tubes they have at the rear end.

A leather-jacket becomes a daddy-long-legs

In September, collect some leather-jackets and put them into soil in a jam-jar. Watch for the change to pupae. Each pupa will arrange itself vertically in the soil so that it just projects from the surface. In about a fortnight, the pupa case splits down the back and the daddy-long-legs pulls itself out.

Earthworms

Earthworms are easy to find. You can turn over a spadeful of moist soil, or look for them on a lawn with a torch at night. They come up to the surface at night to feed on leaves and bits of grass and other plants. During the daytime they stay in burrows in the soil. Sometimes, when there is heavy rain, the burrows become flooded and then you will see many worms come to the surface. Often hundreds of them crawl on to paths. There are several different kinds of worms. If you dig in a compost heap you will find many small red ones. Another kind will plug up the top end of its burrow by pulling a leaf into it. If you see some leaves sticking up vertically in the grass of a lawn, you will know what has caused this.

Looking at earthworms

Bring some worms indoors to look at. You can keep them in a shallow box of moist soil for a time. Put one of the worms on some damp newspaper. What markings can you see on it? Can you see the segments? The under-side of a worm is lighter than

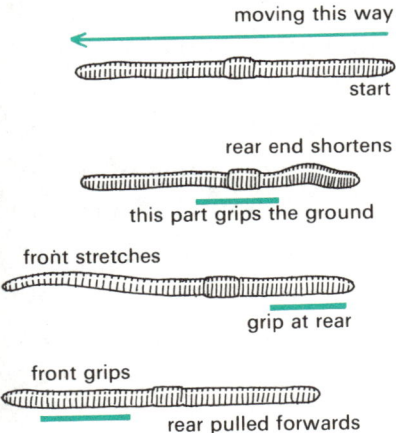

moving this way

start

rear end shortens

this part grips the ground

front stretches

grip at rear

front grips

rear pulled forwards

the top side. Turn one over to look at and leave it to see if it can right itself. Can you see any difference between the head end and the tail end?

Use a magnifying glass to see if you can find the mouth.

How does a worm move?

Put one on a piece of dry brown paper and listen as it crawls along. What do you hear? There are little bristles on the underside of the worm which help it to grip. If you hold up the paper level with your eyes and look against the light from a window, you may just be able to see them. Look with a good magnifier, too.

Put a worm on some damp paper and watch just how it moves forward.

The picture shows what happens.

Making a wormery

Layers of soil kept moist

cover of dark paper to fit jar

Make a 'wormery'. All you need is a transparent container, such as a wide-mouthed jar or a plastic box. Put in some layers of different soils. Each kind should be fine, not hard and lumpy. You could use a sandy soil and a peaty soil. Keep the soils moist but not soaked. Keep the wormery in a cool place, and tape black paper or thin card round it to keep out the light from the sides. Do not put in too many worms; a large jam-jar will take three or four.

Slide up the cover sometimes to see if the worms are making tunnels near the sides. How quickly do they mix up the different layers? Worms burrow through the soil by pushing it to one side and sometimes by eating it. In the soil are tiny bits of old plants. The worms digest these. The rest of the soil

A worm-cast

passes out of the worm's body. Some worms leave this soil underground, but some kinds bring it to the surface and leave little heaps. These heaps are called *worm-casts*. Look for them on lawns and paths. Worms do the soil a lot of good by mixing it up and letting air into it.

The famous scientist, Charles Darwin, watched worms a lot. He estimated that each one brought over 500 grams of soil to the surface in a year, and that there were about 13 worms to the square metre in a garden. Find the area of a patch of lawn or a flower-bed in your garden or at school. What weight of soil do the worms bring to the surface on that patch in a year if Darwin's figures are right?

What do worms like best to eat?

Put small bits of different foods on top of the soil in your wormery. See if any get eaten or pulled into the soil. Try bits of fresh leaves, decaying leaves, carrot, onion, celery, apple peel, bread, etc. The leaves could be lettuce or cabbage, but try different tree leaves, too. Is it true that worms like beech leaves better than others? Moisten a beech leaf and one or two other kinds and put them on top of the soil. Keep the wormery cool and dark for a week. Then look to see what has happened.

Worms and light

For this experiment you will need a darkened room, or maybe you can make a dark space to work in by pinning black cloth or card to the edge of a table so that it reaches to the floor all round.

Worms have no eyes, yet they can sense light. Make a hole in a piece of card and tape the card to the front of a torch so that you can make a narrow beam of light. Place a worm on wet newspaper in the dark. Let it rest for a while and then shine the light on the rear end, the sides and the front in turn. How does the worm respond?

Put a piece of red cellophane over the front of the torch. Does the worm 'feel' red light? Try other colours, too.

hole about 5 mm diameter

Worms and sound

A worm cannot hear but it can feel vibrations. Place a worm on a dish of moist soil on top of a piano. After a while, sound a low note loudly. Does the worm respond?

Earwigs

Earwigs are fully grown in August. They nibble flower petals, so gardeners do not like them very much, but they are harmless creatures really. They feed at night, mainly on small caterpillars, flies and other soft-bodied insects. During the day they press themselves tightly into cracks and corners. Look for them in between flower petals, in inverted flowerpots, under bark and fallen leaves. Sometimes they get into holes in windfall apples. You can collect them by shaking dahlia flowers over a box.

Put some earwigs in different shaped jars and boxes and stand these in a dark place. After a while, look where the earwigs have moved to inside the containers.

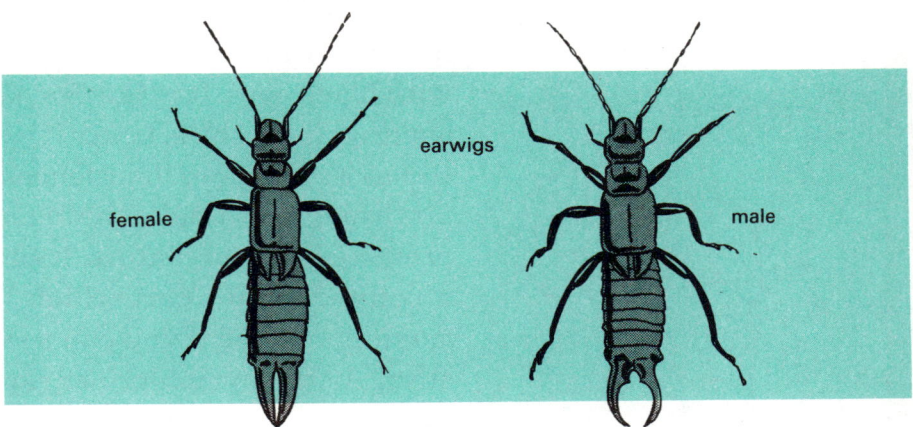

female   earwigs   male

Male and female earwigs

Sort out some males and some females. You can tell which is which by looking at the shape of their 'nippers'. The males have more curved ones. These nippers look fierce but are harmless.

Try them with your finger!

In February and March, the female earwig lays about 20 eggs in soil or leaf litter, and then guards them. She goes on looking after the young earwigs for a while after they are hatched. It is very unusual for an insect to do this.

An earwig's wings

Earwigs have four wings. The front ones are hardened, and are simply covers for the back ones, which are almost always folded up and tucked underneath. When they are opened out they are shaped rather like ears, so perhaps that is how the insect got its name. Very few people have ever seen an earwig fly.

Earwigs have wings!

Cuckoo-spit on plant stem

Froghoppers

In May and June, you will be able to find blobs of white froth on the stems of many plants. Look on grasses, thistles, garden pinks and the young twigs of willow trees. Because this is the time of the year when the cuckoo is heard, the blobs are nicknamed 'cuckoo-spit' but they have nothing to do with the cuckoo.

A cuckoo-spit survey

In May or June, make a 'cuckoo-spit survey' in your district. See what kind of plants get it, and whether one kind of plant has more cuckoo-spit blobs than any other.

On what part of the plants is the cuckoo-spit found?

Take a stiff grass stem and stroke away the froth. You will find inside a small soft creature, pale green or yellow in colour, with black eyes. It is the *nymph* of a froghopper. This animal is making the bubbles. Leave it in position to see how long it takes to make some more. It produces a liquid from its body and blows air through it to make the foam. Inside, the nymph is moist and protected from predators. It sucks sap out of the plant through a pointed mouth part which looks like a beak. Look with a magnifier to see if you can see the 'beak' stuck into the plant stem. As the nymph grows rapidly, it casts its skin from time to time.

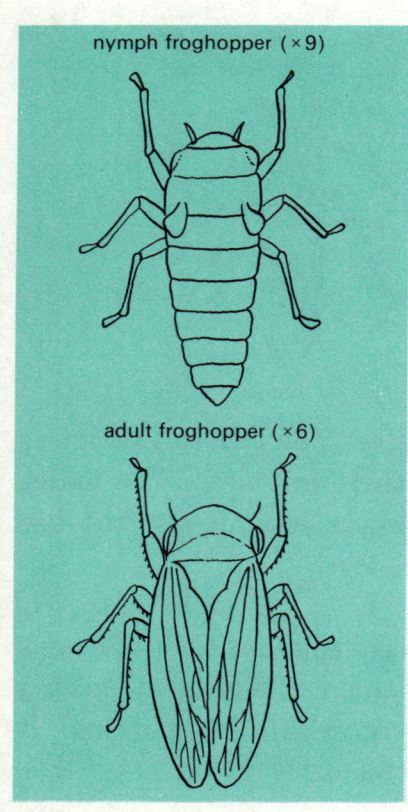
nymph froghopper (×9)

adult froghopper (×6)

What the nymph becomes

The picture also shows an adult froghopper. This is what the cuckoo-spit nymph changes to. It gets its name because it is shaped rather like a frog, and it hops large distances. Sometimes, as you walk along through grass, you will notice numbers of them jumping away from your feet.

Pick a stem with cuckoo-spit on it. Put the stem in water and cover it with a large jar turned upside down. In time, do you get a froghopper?

There are several kinds of froghoppers. Look out for a bright red and black one which lives on willow trees. The commonest one is greyish-brown.

Aphids

Have you seen greenflies on rose-bushes and blackflies on broad bean plants? These insects are *aphids*. There are aphids

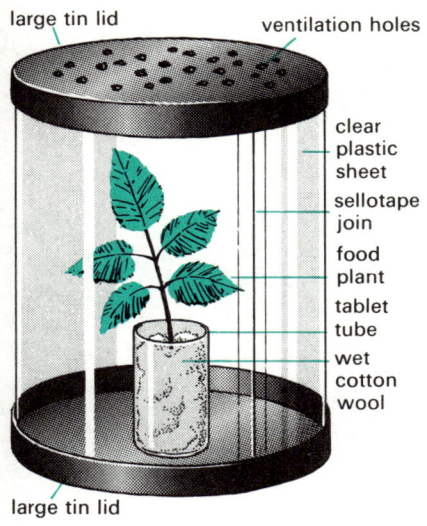

large tin lid

ventilation holes

clear plastic sheet

sellotape join

food plant

tablet tube

wet cotton wool

large tin lid

of other colours, too, but the structure and way of living is the same for all of them. Look for aphids on nettles, fruit trees, lime-trees, sycamore trees and hawthorn bushes. How many different plants can you find with aphids on?

Watching greenflies

Find a rose-bush which has greenflies on it. Cut a shoot and keep it in an insect cage as shown in the picture.

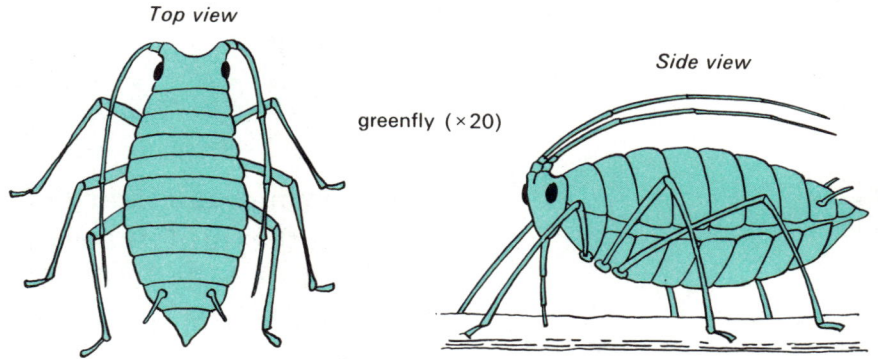

Top view

greenfly (×20)

Side view

Use a magnifying glass to make large drawings of greenflies. Have they got any wings? Have they *all* got wings? Look at their eyes and their mouthparts. Can you see the 'beak' stuck in the plant stem? There are so many crowded together, because each aphid can produce several young ones every day! Can you see any young ones?

How quickly do greenflies reproduce?

Cut another rose-shoot without any aphids on it. Wash it carefully to be sure there are no aphids on it.

Put it in water and transfer *one* fat greenfly on to it with a paintbrush.

Count how many there are after 24 hours, 48 hours, and so on.

Honeydew

Watch your aphids to see if they give out any tiny blobs of 'honeydew' from their rear ends. This is a clear sugary liquid. The aphid usually jerks off the drop. Sometimes there are so many aphids on the leaves of lime-trees and sycamore trees that it 'rains' honeydew underneath. You may have noticed this, especially on cars parked under the trees.

Look at plants with aphids on them to see if there are any ants amongst them. The ants are not eating the aphids. They are coming for the honeydew which the aphids allow them to drink.

Lacewings

These pretty insects with green transparent wings and yellow eyes come into houses in the autumn. Do not kill them because they eat aphids which are garden pests.

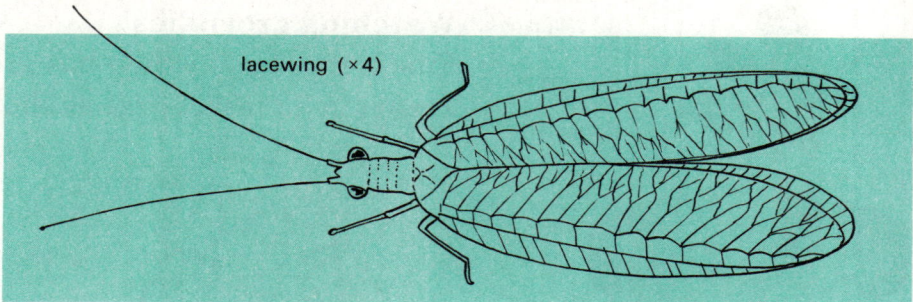

lacewing (×4)

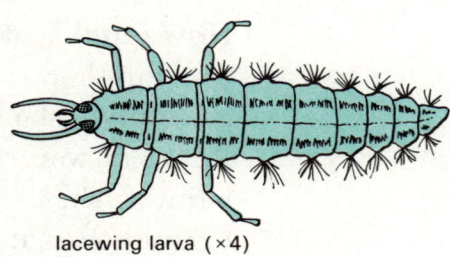

lacewing eggs on twig (×4)

The lacewing has a strange way of laying its eggs. It touches a twig with the tip of its body and gives out a drop of fluid. Then it lifts up its body to draw out the fluid into a thread. The thread hardens in the air, and the lacewing lays one egg and sticks it to the end of the thread. You may be able to find a group of these eggs on a rose-twig. The larva which hatches from one of the eggs is not at all like the lacewing fly.

It has sharp pointed jaws and eats lots of aphids. When it is full-grown it rolls up into a ball, spins a silk cocoon around itself and changes to a pupa inside. After a time, the dainty lacewing fly bites its way out of the cocoon.

lacewing larva (×4)

Index